Kai-Daniel Frank Büchter

Nonlinear Optical Frequency Conversion to and from the Mid-Infrared

Kai-Daniel Frank Büchter

Nonlinear Optical Frequency Conversion to and from the Mid-Infrared

in Ti: PPLN Waveguides for Spectroscopy and Free-Space Optical Communication

Südwestdeutscher Verlag für Hochschulschriften

Impressum/Imprint (nur für Deutschland/only for Germany)
Bibliografische Information der Deutschen Nationalbibliothek: Die Deutsche Nationalbibliothek verzeichnet diese Publikation in der Deutschen Nationalbibliografie; detaillierte bibliografische Daten sind im Internet über http://dnb.d-nb.de abrufbar.

Alle in diesem Buch genannten Marken und Produktnamen unterliegen warenzeichen-, marken- oder patentrechtlichem Schutz bzw. sind Warenzeichen oder eingetragene Warenzeichen der jeweiligen Inhaber. Die Wiedergabe von Marken, Produktnamen, Gebrauchsnamen, Handelsnamen, Warenbezeichnungen u.s.w. in diesem Werk berechtigt auch ohne besondere Kennzeichnung nicht zu der Annahme, dass solche Namen im Sinne der Warenzeichen- und Markenschutzgesetzgebung als frei zu betrachten wären und daher von jedermann benutzt werden dürften.

Verlag: Südwestdeutscher Verlag für Hochschulschriften GmbH & Co. KG
Dudweiler Landstr. 99, 66123 Saarbrücken, Deutschland
Telefon +49 681 37 20 271-1, Telefax +49 681 37 20 271-0
Email: info@svh-verlag.de

Approved by: Paderborn, Universität, Diss., 2011

Herstellung in Deutschland:
Schaltungsdienst Lange o.H.G., Berlin
Books on Demand GmbH, Norderstedt
Reha GmbH, Saarbrücken
Amazon Distribution GmbH, Leipzig
ISBN: 978-3-8381-2993-8

Imprint (only for USA, GB)
Bibliographic information published by the Deutsche Nationalbibliothek: The Deutsche Nationalbibliothek lists this publication in the Deutsche Nationalbibliografie; detailed bibliographic data are available in the Internet at http://dnb.d-nb.de.

Any brand names and product names mentioned in this book are subject to trademark, brand or patent protection and are trademarks or registered trademarks of their respective holders. The use of brand names, product names, common names, trade names, product descriptions etc. even without a particular marking in this works is in no way to be construed to mean that such names may be regarded as unrestricted in respect of trademark and brand protection legislation and could thus be used by anyone.

Publisher: Südwestdeutscher Verlag für Hochschulschriften GmbH & Co. KG
Dudweiler Landstr. 99, 66123 Saarbrücken, Germany
Phone +49 681 37 20 271-1, Fax +49 681 37 20 271-0
Email: info@svh-verlag.de

Printed in the U.S.A.
Printed in the U.K. by (see last page)
ISBN: 978-3-8381-2993-8

Copyright © 2011 by the author and Südwestdeutscher Verlag für Hochschulschriften GmbH & Co. KG and licensors
All rights reserved. Saarbrücken 2011

Contents

Introduction   **1**

1. Strip Waveguides for Nonlinear Frequency Conversion   **7**
    1.1. Theoretical Treatment of Periodically Poled Waveguides 8
        1.1.1. Waveguide and Non-linear Optical Polarization Theory 8
        1.1.2. Coupled Mode Theory 11
        1.1.3. Quasi Phase-Matching 15
    1.2. Waveguide Sample Fabrication 16
        1.2.1. Waveguide Fabrication 17
        1.2.2. Periodic Poling 18
    1.3. Summary and Discussion 19

2. Mid-infrared Source based on Difference-Frequency Generation   **21**
    2.1. Theoretical Modelling 22
    2.2. Experimental Setup 24
    2.3. Sample Characterization & Discussion of Results 28
        2.3.1. Mid-infrared Mode Distributions 28
        2.3.2. Nonlinear Conversion Efficiency 30
        2.3.3. Wavelength Characteristics 34
        2.3.4. Output-Power Stability 35
    2.4. Summary 36

3. Hybrid Up-Conversion Detector   **38**
    3.1. Theoretical Background 39
        3.1.1. Basic Detector Figures of Merit 39
        3.1.2. Up-conversion Detector Performance Estimation 40
    3.2. Up-Conversion Detector based on Sum-Frequency Generation 46
        3.2.1. Experimental Setup 46
        3.2.2. Results & Discussion of Detector Performance 48
    3.3. Up-conversion Detector based on Difference-Frequency Generation 54
        3.3.1. Experimental Setup 54
        3.3.2. Results & Discussion of Detector Performance 55
    3.4. Additional Considerations 57
    3.5. Summary 59

4. Absorption Spectroscopy using Frequency Conversion   **61**
    4.1. Basics of Trace Gas Spectroscopy 62

4.2.	Absorption Spectroscopy using DFG-MIR Source	64
4.3.	Absorption Spectroscopy using DFG-MIR Source and SFG-UCD	66
4.4.	Frequency Modulation Spectroscopy using DFG Source and SFG-UCD	68
4.5.	Summary	68

5. Free-space Optical Transmission in the MIR using Wavelength Conversion 72
 5.1. Atmospheric Transmission Impairments 74
 5.2. Down- and Up-Conversion for Free-Space Optical Transmission 76
  5.2.1. Experimental Setup 77
  5.2.2. Free-Space Transmission-Line Characteristics 78
 5.3. Evaluation of Free-Space Data-Transmission 84
 5.4. Summary 87

Conclusions 88

A. Waveguide and Poling Mask Design 92

B. Modeling of Lens Dispersion 95

C. Waveguide Inhomogeneity Calculations 97

D. Reference HgCdTe Detector Characterization 99

E. Atmospheric Transmission Impairments 101
 E.1. Atmospheric Scattering and Absorption 101
 E.2. Beam Divergence and Spreading 105
 E.3. Scintillation 106

List of Figures

0.1. SFG / DFG energy level diagram. 2
0.2. Overview of MIR lasers. 4
0.3. Free-space transmission using frequency conversion. 5

1.1. Illustration of Ti-indiffused waveguide in $LiNbO_3$. 8
1.2. Fundamental electric field distributions in an MIR-waveguide. 11
1.3. SFG and DFG efficiency calculations. 14
1.4. Nonlinear-optic generation of radiation over short interaction lengths. . . 16
1.5. Waveguide fabrication steps. 17
1.6. Periodic poling steps. 18

2.1. Calculated idler power dependence on pump and signal powers. 23
2.2. DFG-MIR source scheme and photo. 25
2.3. Illustration of lens dispersion. 25
2.4. DFG setup and fiber coupling details. 26
2.5. Transmission properties of input anti-reflection coating. 27
2.6. MIR waveguide modes. 29
2.7. Measured and calculated Ti-waveguide MIR mode profiles. 29
2.8. DFG tuning characteristics. 30
2.9. DFG power characteristics. 32
2.10. Senza pump laser. 33
2.11. Two-dimensional DFG power plot. 33
2.12. DFG phase-matching temperature dependence. 34
2.13. DFG stability measurements. 36

3.1. Principle of SFG and DFG up-conversion detectors. 38
3.2. Ideal detectivity of photo-conductive / photo-voltaic detectors. 42
3.3. NEP of shot-noise / background limited detectors. 45
3.4. Calculated UCD conversion efficiencies, including losses. 46
3.5. Experimental setup for UCD characterization. 47
3.6. SFG up-conversion detector scheme. 48
3.7. Femto OEC-200-IN2 noise characteristics. 49
3.8. SFG up-conversion detector characterization setup. 50
3.9. Up-conversion detector pump reflection mirror characteristics. 51
3.10. SFG up-conversion detector power characteristics. 51
3.11. Direct comparison of MCT and SFG-UCD. 52
3.12. Spectrum of SFG up-conversion detector output. 53
3.13. DFG up-conversion detector scheme. 55

3.14. DFG up-conversion detector characterization setup. 55
3.15. DFG up-conversion detector power characteristics. 56
3.16. Parametric gain in DFG up-conversion detector. 57
3.17. Parametric fluorescence spectrum of the DFG-UCD at 1550 nm. 58

4.1. Absorption spectroscopy setup using conventional detection. 63
4.2. Absorption measurement using DFG-MIR source and a MCT detector. . 64
4.3. Normalized absorption measurement using DFG source. 65
4.4. Absorption spectroscopy using SFG-UCD. 66
4.5. Photo of DFG-SFG absorption setup. 67
4.6. Methane absorption measurement using DFG and SFG. 67
4.7. FM spectroscopy using SFG-UCD. 69
4.8. Frequency modulation spectrum of methane. 69

5.1. Illustration of free-space optical links. 72
5.2. Wavelength dependence of atmospheric attenuation. 75
5.3. Free-space transmission using frequency conversion. 76
5.4. Power characteristics of the FSO transmitter module. 77
5.5. DFG power characterization setup for FSO transmission modules. 78
5.6. FSO transmission experiment using nonlinear wavelength conversion. . . 79
5.7. Pump-reflection mirror of the receiver sample. 79
5.8. Photo of Transmitter module. 80
5.9. Transmitter and Receiver tuning characteristics. 81
5.10. Transmitter power characteristics. 82
5.11. FSO-transmission line characteristics. 83
5.12. Losses in the FSO transmission line. 83
5.13. FSO data transmission setup. 85
5.14. BER impairment with wavelength converters. 86
5.15. Free-space transmission line through a wind tunnel. 87

A.1. Waveguide group layout. 92
A.2. Nonlinear efficiency (DFG) as function of strip width and thickness. . . . 93
A.3. Mask layout with six poling groups. 94

B.1. Gaussian beam modeling of lens dispersion. 96

C.1. Phase-matching chirp assumed for modeling. 98

D.1. MCT detector characterization. 99

E.1. Molecular and Aerosol scattering and absorption coefficients. 102
E.2. Atmospheric transmission considering absorption losses. 103
E.3. High resolution atmospheric transmission spectra. 104
E.4. Atmospheric transmission considering scattering and absorption losses. . 105
E.5. Scintillation index for weak, moderate, and strong turbulence conditions. 108

List of Tables

2.1. Effect of waveguide inhomogeneities on $\Delta\beta$. 31

3.1. Comparison of theoretical detectivities at different wavelengths. 44

E.1. Transmission bands in the near- to mid-infrared range. 104

List of Abbreviations

AC	Alternating Current
AFW	Australian Fibre Works
APD	Avalanche Photo Diode
AR	Anti Reflection
ARC	Anti Reflection Coating
BER	Bit Error Rate
BERT	Bit Error Rate Tester
bg	Background
BLIP	Background Limited Performance
BPF	Band Pass Filter
C-band (optical)	Wavelength range from 1530-1565 nm; corresponding to EDFA amplification range
CME	Coupled Mode Equations
cw	Continuous Wave
DC	Direct Current
DF	Difference Frequency
DFG	Difference Frequency Generation
DPSS	Diode Pumped Solid State (- laser)
E-beam	Electron-beam
ECL	External Cavity Laser
EDFA	Erbium Doped Fiber Amplifier
EDTA	Ethylenediaminetetraacetic acid
FBG	Fiber Bragg Grating
FC	Fiber coupling (also: flat polish connector in fiber optics)
FEM	Finite Element Method
FSO	Free-Space Optical
FTIR	Fourier Transform Infrared
GEISA	Gestion et Etude des Informations Spectroscopiques Atmosphériques (Management and Study of Spectroscopic Information) (spectroscopic database)
HD	Hybrid Detector
HITRAN	HIgh-resolution TRANsmission molecular absorption database (spectroscopic database)
HWP	Half Wave Plate
ICAO	International Civil Aviation Organization
IEEE	Institute of Electrical and Electronics Engineers
IR	Infrared

ISO	(optical) Isolator
ITU	International Telecommunications Organization
LAN	Local Area Network
LASER	Light Amplification by Stimulated Emission of Radiation
LD	Laser Diode
LN	$LiNbO_3$ / Lithium Niobate
LPF	Low-pass Filter
MCT	Mercury Cadmium Telluride (HgCdTe)
MIR	Mid-Infrared
Mod.	Modulation
NIR	Near-Infrared
NRZ	Non-Return to Zero
OAP	Off-axis Paraboloid mirror
OCT	Optical Coherence Tomography
OPO	Optical Parametric Oscillation / Oscillator
OSA	Optical Spectrum Analyzer
PBS	Polarization Beam Splitter
PbXXXz	Waveguide numbering format (Paderborn, Sample Number, z for Z-cut)
PM	Polarization Maintaining (fiber)
PPLN	Periodically Poled Lithium Niobate
QCL	Quantum Cascade Laser
QKD	Quantum Key Distribution
QPM	Quasi Phase Matching
QPSK	Quadrature Phase-Shift Keying
QTE	Quasi Transversal Electric
QTM	Quasi Transversal Magnetic
QW	Quantum Well
QWP	Quarter Wave Plate
RT	Room Temperature
RX	Receiver
SF	Sum Frequency
SFG	Sum Frequency Generation
SH	Second Harmonic
SHG	Second Harmonic Generation
Sig.	Signal
SPECTRA	Spectroscopy of Atmospheric Gases (spectral calculation software)
TE	Transversal Electric
TM	Transversal Magnetic
TX	Transmitter
UC	Up-Converter
UCD	Up-Conversion Detector
UV	Ultra Violet
VMR	Volume Mixing Ratio

WDM .... Wavelength Division Multiplexing
WG .... Waveguide
WLAN .... Wireless Local Area Network

Introduction

The current year 2010 marks the 50th anniversary of the LASER (Light Amplification by Stimulated Emission of Radiation), which was first demonstrated in an experiment by Maiman in 1960 [1], following the proposal by Schawlow and Townes to realize an Infrared MASER [2]. Since then, the laser has proven to be an invaluable asset to our daily lives. Lasers provide the backbone to numerous industries: they play major roles in telecommunications, in fabrication processes, in optical storage, in aerosol and gas detection; e.g. for combustion diagnostics and trace gas monitoring, in medical applications; e.g. surgical lasers and optical coherence tomography (OCT), etc. Lasers generate coherent radiation with narrow spectral linewidths and high spectral intensities, permitting access to hitherto unavailable physical phenomena.

The invention of the laser also enabled optical frequency conversion, which was first demonstrated by Bloembergen et al. [3] using nonlinear optical crystals. The potential of nonlinear optical frequency conversion using lasers is quite extraordinary, due to the fact that this technique allows us to access wavelength ranges which are impractical to reach using, for example, diode lasers. Moreover, it was proposed early on to use frequency conversion for low-light detection of infrared radiation [4], using visible light detectors.

To put it more generally, frequency (or wavelength) conversion enables us to use radiation at an arbitrary wavelength for a specific purpose, while the instrumentation used may operate at a different wavelength. As an example, a frequency (or wavelength) converter could act as an interfacing node between two optical communication networks, one operating at a longer wavelength (e.g. for long-haul distribution using silica fibers), and one operating at a shorter one (e.g. for fiber-to-the-home applications using polymer optical fibers).

Due to the impressive prospects, nonlinear optical frequency conversion was extensively studied throughout the last decades. The basic physical property enabling the conversion is the dielectric polarization of a given material, which in principle allows the mixing of arbitrary frequencies to generate new ones. The mixing itself is due to the nonlinearity of the polarization field when high electric field strengths are present.

The most widely used nonlinear optical phenomenon is Second-Harmonic Generation (SHG), where the frequency of a pump laser is doubled in a nonlinear optical crystal. The attractiveness of this concept is illustrated if we consider that efficient diode-pumped solid-state (DPSS) lasers are available in the near infrared, and doubling may lead to the generation of green light. The laser radiation in this case is called fundamental wave, and the generated green light is the second harmonic wave. While cheap diode lasers producing red light emission are widely available, the human eye is more sensitive to the green, so the concept of frequency-doubled infrared lasers can be found in green laser pointers.

The concept can be taken further by adding an additional source of radiation with a different wavelength. The radiation at the two wavelengths is usually designated as pump and signal. If we consider first-order nonlinear interactions, the superposition of both waves will generate additional frequency components at the sum and difference frequencies, respectively. In this way, wavelengths may be converted almost arbitrarily, given a nonlinear optical crystal which is transparent at the wavelengths involved. In the photon-picture, the process may be illustrated by introducing virtual energy levels (figure 0.1). Based on the law of energy conservation, the incident pump and signal photons can combine to generate a new photon, causing annihilation of the two incident photons. Otherwise, the incident pump photon may also decay to generate two new photons, with one having the exact wavelength of the incident signal photon. In the first case we talk about Sum Frequency Generation (SFG), in the second case of Difference Frequency Generation (DFG). Wavelength tuning can be done by tuning either the pump or signal sources.

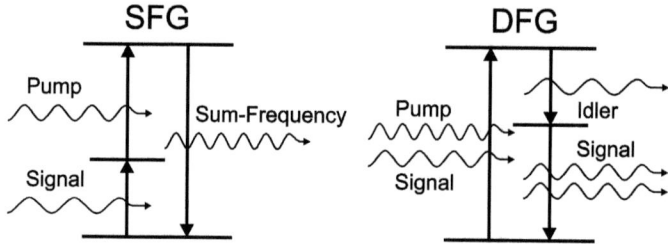

Figure 0.1.: Virtual energy level diagrams for Sum- (SFG) and Difference- Frequency Generation (DFG). In the case of SFG, two incident photons (in the photon-picture) combine to generate a new photon at the sum-frequency. In the case of DFG, a pump-photon decays into two photons: one with the energy of the incident signal photon; one with the energy difference of pump and signal. The initial signal photon is preserved.

Due to dispersion, optical waves at different frequencies (or different wavelengths, respectively), are subject to unequal refractive indices and thus unequal phase velocities. As a consequence, the dielectric polarization oscillation at the newly generated frequency must be phase-matched to a propagating wave at the same wavelength for an efficient energy transfer to the newly generated wave. Nowadays, the most widely used approach to achieve this is called quasi-phasematching (QPM) by periodic poling, which was proposed by Bloembergen et al. [3] and demonstrated in lithium niobate (LiNbO$_3$) waveguides in the early 1990ies by Jundt and others [5].

From a technical point of view, there are two competing approaches in optics for nonlinear devices: the more conventional approach is to use bulk optics, where laser radiation in the form of Gaussian beams is shone through a crystal. In this case, the nonlinear efficiency is governed by crystal length and beam parameters. The (arguably) more sophisticated approach is to use waveguiding structures, where radiation is coupled to an optical mode in a waveguide by means of a lens or a fiber, and guided wave interactions are exploited. The tightly confined mode is free of divergence and can be maintained over long interaction lengths. In the case of a waveguide, mode dispersion needs to be considered in addition to material dispersion.

Within the scope of this work, nonlinear frequency conversion in optical waveguides was exploited to generate mid-infrared radiation from near-infrared radiation, as well as to re-generate near-infrared radiation from mid-infrared radiation. Here, we define as near-infrared the wavelength-range starting from 750 nm at the edge of the visible spectrum, up to about 2.5 μm, where the mid-infrared starts. More specifically, wavelengths in the 1- to 1.5-μm and 3- to 4-μm wavelength ranges were used.

For the purpose of wavelength conversion, we developed modules exploiting wavelength down- and up-conversion in Ti-indiffused waveguides in periodically poled lithium niobate (Ti:PPLN). The motivation for this approach to access the mid-infrared wavelength range is manifold: On the one hand, mid-infrared radiation sources are invaluable tools for vibrational spectroscopy of trace gases. They are of interest in free-space optical communications for intermediate distances, and medical applications due to strong water absorption [6]. However, mid-infrared lasers with certain properties like narrow linewidth and wavelength tunability have in the past been unavailable for certain wavelength ranges. An overview of current existing MIR lasers is shown in figure 0.2 (color center lasers are omitted) - only lead-salt lasers as well as quantum cascade lasers have reached commercial status and seem widely applicable; however both have certain restrictions (lead-salt lasers exhibit, for example, strong mode-hopping behavior [7]. QCL's, on the other hand, offer only limited wavelength coverage and can have high thresholds [8]). Solid state lasers are also available, however only at fixed wavelengths. In consequence, Difference Frequency Generation (DFG) and Optical Parametric Oscillation (OPO) in nonlinear optical crystals have been a popular means to realize narrow linewidth, wavelength-tunable mid-infrared lasers [9, 10, 11].

Also, while considerable research was done concerned with the fabrication of high-quality mid-infrared detectors, they suffer from intrinsic shortcomings, in part due to the fact that they are sensitive to thermal radiation. On the other hand, near-infrared diode lasers, photo detectors, as well as fiber-optic components operating in the near-infrared exhibit excellent properties, due to the economic driving force of the optical communications industry. Combining both, nonlinear wavelength conversion and well-developed near-infrared detectors, can thus be an interesting alternative. In addition, diverse NIR components, e.g. for data encoding and decoding, can be combined with such an up-conversion detector and circumvent some of the difficulties in mid-infrared applications, for example by allowing increased modulation bandwidths of a signal.

In essence, two devices were developed for frequency conversion to and from the mid-infrared within the framework of this thesis:

- Firstly, a fiber-coupled MIR-generator (Transmitter) using wavelength conversion

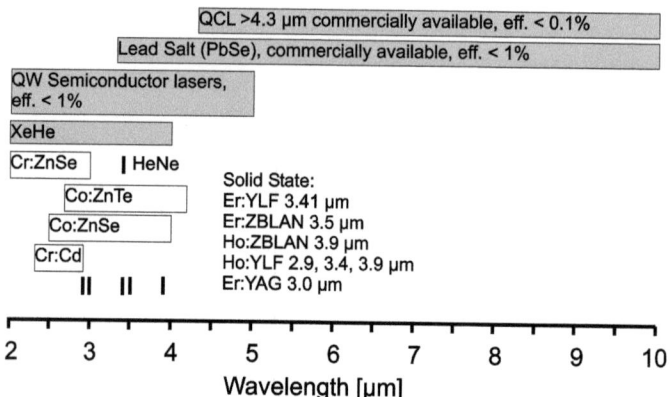

Figure 0.2.: Overview of MIR lasers, compiled from [9, 8]. Mainly lead-salt lasers and QCL's are compact, robust, tunable, as well as commercially available. QW: Quantum Well

from the NIR to the MIR was built. It is shown that high conversion efficiencies can be attained and that amplitude as well as phase information of the NIR waves are preserved during the wavelength conversion.

- Secondly, a hybrid up-conversion detector (Receiver) is demonstrated, which converts MIR radiation to the NIR. The advantage of such a detector lies in a potential increase in sensitivity compared to a conventional MIR detector, due to superior properties of NIR detectors. Also, as phase information is preserved, it may be used together with phase-sensitive instrumentation, e.g. QPSK (Quadrature Phase Shift Keying) systems.

The MIR-generator and up-conversion detector used in this work are essentially identical in construction, therefore in principle, functionality can be exchanged. The wavelength converters employ Ti-indiffused waveguides in $LiNbO_3$ for two reasons: firstly, optical waveguides allow for a long interaction length and well-confined field distributions, which is both crucial for an efficient wavelength conversion. Secondly, waveguides are compatible to optical fibers, in the sense that radiation may be coupled directly between optical fibers and waveguides (given suitable wavelengths). In that way, efficient and compact packaged modules, which can easily be interfaced with standard fiber optic devices, are demonstrated. External efficiencies in excess of 10 %/Watt achieved and output powers above 10 milliwatts were achieved.

It is shown that the hybrid detector can be designed to be more sensitive than a Mercury Cadmium Telluride (HgCdTe or MCT) detector optimized for 3.4-μm wavelength, if an EDFA (erbium doped fiber amplifier) is used to boost the pump, together with adequate pump filtering. In part, the superior performance is achieved due to the highly wavelength-selective up-conversion. In addition, it is shown that the hybrid detector can convert sufficient power even without pump amplification, e.g. for spectroscopic applications using conventional fiber-optics and wavelength conversion. Using both, transmitter and receiver modules, a transmission line is demonstrated with external (i.e., not waveguide internal) pump/signal powers in the low milliwatt-range, to demonstrate MIR spectroscopy using NIR components only.

A large part of the results presented here stems from a collaboration with the University of Stanford, CA, which supported the University of Paderborn as a subcontract of CeLight, Inc. Within that framework, an NIR-MIR-NIR transmission line (figure 0.3) was used to transmit a QPSK modulated signal over a free-space MIR link. Nonlinear down- and up-conversion was used to convert a phase-coded signal from 1550 nm to 3.8 µm and back to 1550 nm. The project succeeded in showing that free-space transmission at 3.8 µm is less susceptible to certain atmospheric mitigation effects than 1550 nm radiation. Especially Carsten Langrock, Ph.D. (University of Stanford, CA) has contributed to the successful outcome of this project. References to associated works, conducted together with Ezra Ip, Ph.D. (University of Stanford, CA), as well as project associates of CeLight, Inc. and The Boeing Company, are given in the appropriate chapters.

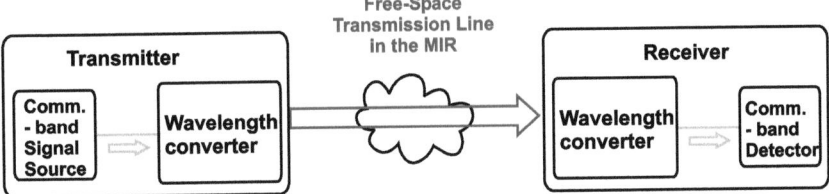

Figure 0.3.: Free-space optical transmission line in the mid-IR using frequency conversion. The mid-IR wavelength may be chosen nearly arbitrarily, using appropriate wavelength converters, to mitigate atmospheric transmission impairments. A similar arrangement can be used e.g. for optical absorption measurements in the mid-IR.

Some of the research within the scope of this dissertation is built on previous work, conducted during my time as a M.Sc. candidate [12], which was concerned purely with the application of Ti:PPLN waveguides for mid-infrared generation for spectroscopy on methane (CH_4). Also, the work conducted within the scope of the B.Sc.-thesis by Marie-Christin Wiegand, née Fischer, has led to some results presented in this dissertation [13]. I have structured this thesis as follows:

- In the first chapter, a review on modeling of nonlinear optical interactions in waveguides, and an overview of the fabrication of periodically poled $LiNbO_3$ waveguides, are presented.

- In the second chapter, near-infrared pumped mid-infrared sources using DFG are discussed, building on the previous chapter, and experimental results are presented.

- In the third chapter, hybrid up-conversion detectors, using both SFG and DFG with near-infrared pump lasers and detectors, are discussed with regard to conventional mid-infrared detectors.

- In the fourth and fifth chapters, free-space transmission in the mid-infrared, using the DFG source and SFG/DFG based up-conversion detectors, is discussed. The application to trace-gas spectroscopy and free-space communications is demonstrated.

- Lastly, a summary is given.

Each chapter begins with a short introduction and ends with a summary. I have tried to present sufficient theoretical background at the beginning of each chapter, in order to motivate the developments and investigations. Bear in mind that the theoretical treatments are not meant to be exhaustive, as each field (waveguide modeling, detector modeling, molecular absorption modeling, and atmospheric transmission modeling) is complex in itself. In addition, I have moved some more theoretical discussion to the appendix, in order to keep the main chapters short and consistent, and to keep the focus on the achieved results.

1. Strip Waveguides for Nonlinear Frequency Conversion

Waveguides are the main building blocks of integrated optical devices. The field of integrated optics may be understood as the optical equivalent to integrated electronic circuits, where electronic functionalities are combined on a common substrate (e.g. a Silicon chip). However, there is no dominant material like Silicon in electronics, but a plethora of different substrates are available; each with different strengths and weaknesses. Depending on the intended application, semiconductors; glasses, polymers, or dielectric crystals may be favored [14].

For frequency conversion purposes, ferroelectric crystals like $LiNbO_3$ are very attractive candidates. $LiNbO_3$ is an artificial crystal that exhibits a strong optical nonlinearity and may be periodically poled. Both features are important for efficient frequency conversion devices. In addition, it supports low-loss optical waveguides, for example by Ti-indiffusion. While bulk optic crystals are convenient means to offer frequency conversion, the integrated optical approach has certain advantages; the first and foremost being that light is confined to small cross-sections within the waveguide in discrete guided modes. In effect, high field intensities are intrinsically provided and may be maintained over long interaction lengths. In the bulk optical case, the effective interaction length is restricted to the Rayleigh range of interacting Gaussian beams. Due to wavelength dispersion in optical crystals, as well as modal dispersion in waveguides, a method to compensate any phase mismatch of interacting light waves is generally necessary. This is either achieved by birefringent phasematching or by quasi-phasematching by using a periodically poled crystal.

In this chapter, a short introduction to integrated optics and especially Ti-indiffused optical waveguides in $LiNbO_3$ is given. In the first sub-chapter, a recapitulation of the theoretical treatment of optical mode propagation and nonlinear interactions in such waveguides is presented, while the second sub-chapter elaborates on the waveguide fabrication process in some detail. For a more accurate description of fabrication and theoretical description of mid-infrared waveguides in Ti:PPLN, see e.g. [9, 10]. At the end of this chapter, the distinctions between sum-frequency and difference-frequency generation are discussed with respect to realizing a hybrid up-conversion detector.

1.1. Theoretical Treatment of Periodically Poled Waveguides

In order to describe nonlinear optical interactions in waveguides, one usually treats waveguide propagation of radiation separately from the actual nonlinear interaction. Radiation propagates in so-called optical modes, each with wavelength- and waveguide-specific electric field distributions, which are calculated from the refractive index distribution within a waveguide cross-section. The nonlinear interaction itself is treated by so-called coupled mode equations. Basic mathematical treatment is introduced in the following.

1.1.1. Waveguide and Non-linear Optical Polarization Theory

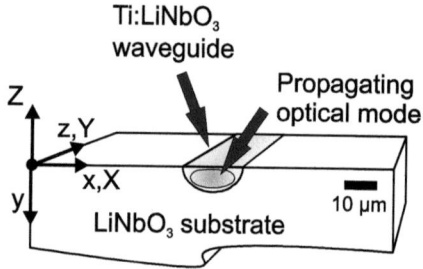

Figure 1.1.: Illustration of Ti-indiffused waveguide in $LiNbO_3$. Upper case letters denote the crystallographic axis; lower case letters the lab coordinates. An approximate scaling bar is given with respect to the dimensions of a typical MIR waveguide.

In figure 1.1, a Ti:PPLN waveguide is illustrated. The Cartesian coordinate system is defined in order to describe electric field evolution, in addition to the crystal coordinate system. An approximate scale bar is given, with respect to the dimensions of a typical Ti-indiffused, MIR-waveguide.

Due to the small dimensions of an optical waveguide, which are roughly comparable to the wavelength of light, propagation within the waveguide is restricted to discrete modes. This is due to the fact that electromagnetic radiation consists of a superposition of oscillating fields, interfering constructively and destructively within a sort of cavity formed by the waveguide. If we consider an optical mode, there are states (the guided modes) where certain boundary conditions are fulfilled and the wave can propagate due to constructive interference within the waveguide. If such conditions are not met, any electromagnetic wave carrying energy will simply emanate from the waveguide. Starting point to describe a waveguide in a charge-free material are Maxwell's equations:

$$\nabla \times \vec{E} = -\frac{\partial \vec{B}}{\partial t} \tag{1.1}$$

$$\nabla \cdot \vec{D} = 0 \tag{1.2}$$

$$\nabla \times \vec{B} = \mu \frac{\partial \vec{D}}{\partial t} \tag{1.3}$$

$$\nabla \cdot \vec{B} = 0, \tag{1.4}$$

with the local electric field vector $\vec{E}$; $\vec{D}$ is the local electric displacement field vector and $\vec{B}$ is the local magnetic field vector. $\mu = \mu_r \mu_0$ is the magnetic permeability; we can let $\mu_r = 1$. Then, the displacement field may be written as

$$\vec{D} = \epsilon \vec{E} = \epsilon_0 \vec{E} + \vec{P}, \tag{1.5}$$

where $\epsilon = \epsilon_r \epsilon_0$ is the permittivity, with relative permittivity ϵ_r. $\vec{P} = \epsilon_0 \underline{\chi} \vec{E}$ is the local polarization of the material, with the susceptibility tensor $\underline{\chi}$. Taking the curl of (1.1) and using (1.2) gives

$$\Delta \vec{E} - \frac{1}{c^2} \frac{\partial^2}{\partial t^2} \vec{E} + \nabla \left[\frac{\vec{E}}{n^2(x,y)} \nabla n^2(x,y) \right] = \mu_0 \frac{\partial^2}{\partial t^2} \vec{P}, \tag{1.6}$$

with the vacuum speed of light $c = (\epsilon_0 \mu_0)^{-\frac{1}{2}}$. The lateral index distribution is described by $n(x,y)$, which is assumed as constant in z for a straight waveguide.

The polarization of the material responds nonlinearly at high field intensities and can be written as a Taylor-series in the following way:

$$\vec{P} = \epsilon_0 \left(\underline{\chi}^{(1)} : \vec{E} + \underline{\chi}^{(2)} : \vec{E}\vec{E} + \underline{\chi}^{(3)} : \vec{E}\vec{E}\vec{E} + ... \right). \tag{1.7}$$

The underlines denote tensor-character of the susceptibility terms. The first term $\underline{\chi}^{(1)}$ is responsible for the linear response of a material. When strong electric fields $\vec{E}$ are present, higher order terms exist which cause additional frequency components to appear. We will restrict discussion to the second order nonlinear term $\underline{\chi}^{(2)}$, which is responsible for frequency mixing effects like second-harmonic generation (SHG), sum frequency generation (SFG), difference frequency generation (DFG) and parametric amplification in case of electric fields at optical frequencies, as well as the linear Pockels-effect in case of additional static electric fields, in non-centrosymmetric media [15]. Further discussion will focus on the two frequency mixing processes SFG and DFG, which can both be used for up-conversion purposes. DFG allows both up- and down-conversion of radiation.

For further considerations we can split the polarization into a linear and a first order nonlinear term

$$\vec{P} = \vec{P}_{lin} + \vec{P}_{nl} \tag{1.8}$$

with $\vec{P}_{lin} = \epsilon_0 \chi^{(1)} : \vec{E}$, representing the linear susceptibility given by $\chi = (\epsilon_r - 1)$, and $\vec{P}_{nl} = \epsilon_0 \chi^{(2)} : \vec{E}\vec{E}$, so equation (1.6) can be written as

$$\Delta \vec{E} - \frac{n^2(x,y)}{c^2} \frac{\partial^2}{\partial t^2} \vec{E} + \nabla \left[\frac{\vec{E}}{n^2(x,y)} \nabla n^2(x,y) \right] = \mu_0 \frac{\partial^2}{\partial t^2} \vec{P}_{nl}. \tag{1.9}$$

If we consider a mode at frequency ω propagating in z-direction, we can take

$$\vec{E} = \frac{1}{2} \left\{ \vec{E}(x,y) e^{i(\omega t - \beta z)} + c.c. \right\}$$

and

$$\vec{P} = \frac{1}{2} \left\{ \vec{P}_{nl}(x,y) e^{i(\omega t - \phi(z,t))} + c.c. \right\},$$

with the complex conjugate c.c.. Here, we assume a z-independent mode field distribution $\vec{E}(x,y)$ and a propagation constant $\beta = n_{eff} k_{vac}$ with the wavevector $k_{vac} = \frac{\omega}{c}$ in vacuum. n_{eff} is the effective index of a waveguide mode at frequency ω. $\phi(z,t)$ is an arbitrary phase factor which depends on the non-linear polarization inducing electric field(s). We may neglect the third addend in equation (1.9) due to the small index gradient in case of a Ti-indiffused waveguide, and arrive at

$$\left[\frac{\partial}{\partial x^2} + \frac{\partial}{\partial y^2} - \beta^2 + n^2(x,y) \frac{\omega^2}{c^2} \right] \vec{E} = -\omega^2 \mu_0 \vec{P}_{nl}, \tag{1.10}$$

which is a time-independent equation for the local electric field of a wave restricted to propagation in z-direction. The homogeneous part on the left hand side of this equation is an eigenvalue-equation in β. It may be used to describe wave propagation of an optical mode by itself, in analogy to a potential well problem in quantum mechanics, if the index distribution $n(x,y)$ is known.

The mode is confined two-dimensionally within the waveguide cross-section and may propagate in the perpendicular direction. It turns out that there are basically two types of orthogonally polarized modes, called quasi-transversal electric (QTE) and quasi-transversal magnetic (QTM) modes. They are called 'quasi'-transversal, because small longitudinal field components exist in the solutions to equation 1.10, in addition to the dominant transversal components. In the following, the terms 'TE/TM' are used instead of 'QTE/QTM'. Depending on the optical frequency and waveguide geometry, a number of modes of different order may exist.

For a fixed wavelength, a waveguide may usually be designed in such a way that only a fundamental TE_{00} or TM_{00} mode exists, which is desired for most practical applications. For the samples used in this work, waveguides were structured in Z-cut $LiNbO_3$ due to technological reasons. In this case, the largest nonlinear coefficient of $LiNbO_3$, d_{33}, is available when TM-polarized modes are used for the nonlinear interaction. The modes must be calculated numerically due to the inhomogeneous diffusion profile. This has been done, for TM-modes, with the finite element method using FOCUS [16, 17]. In figure 1.2, calculated field distributions at 3400 nm, 1550 nm and 1064 nm are depicted.

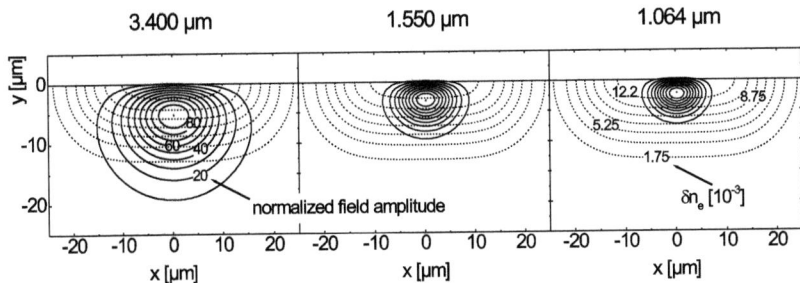

Figure 1.2.: Calculated electric field distributions (solid lines) of fundamental waveguide modes at denoted wavelengths, in an 18-μm wide, Ti-indiffused waveguide cross-section. The (analytically calculated) index variation δn_e of the extraordinary index of LiNbO$_3$ due to Ti-indiffusion is shown as well (broken lines; max. $\delta n_e \approx 14 \cdot 10^{-3}$. The δn_e-profile is slightly wavelength dependent, however the isolines are equivalent). Note the wavelength-dependent position of the modes due to the asymmetric waveguide profile.

Later in this chapter, the effect of field distributions on the nonlinear interaction will be discussed. Higher order waveguide modes and their interactions are not discussed in this thesis, as they were treated extensively in MIR Ti:PPLN waveguides in [12]. In case of the MIR waveguides investigated here, they offer lower conversion efficiency and, in general, less stable and less efficient waveguide coupling.

The nonlinear polarization term on the right hand side of equation 1.10 can be seen as a perturbation and may act as a source term for a wave. In case of a second order nonlinearity due to a $\chi^{(2)}$-interaction according to equation (1.7), optical waves of different wavelengths can mix and generate sum, difference and second harmonic frequencies. In the wave picture, the superposition of two electric fields generates a polarization response with such frequency components (in order to generate a traveling wave at the generated frequency however, phase-matching is generally needed due to material dispersion; this will be discussed in section 1.1.3). In the photon picture, a photon of high energy interacting with a photon of low energy will either generate a new photon with the sum of both energies, or the high energy photon will decay, and two new photons according to the energy difference will be generated. In order to treat nonlinear frequency conversion, coupled mode theory is usually used, which is discussed in the following chapter.

For a more extensive discussion of optical nonlinear effects, consult the relevant literature, e.g. [18].

1.1.2. Coupled Mode Theory

Both, SFG and DFG can be explained starting from the same coupled differential equations, called coupled mode equations (CME). CME were introduced for microwave devices in the 1950's and were later applied to optical waveguides [19]. They can be used to model various kinds of interactions, for example in Ti:PPLN waveguides, ranging from acousto- [20] or electro-optic polarization mode conversion [21], coupling by directional couplers and gratings [22], to non-linear optical interactions discussed here. The equations effectively describe the energy transfer between a number of electric fields and give accurate results when the assumption of a finite number of discrete modes is valid. In waveguides with low mode indices, this assumption is reasonable.

In the following, the slowly varying field amplitudes $A_{m,n}^{(\omega)}(z)$, at frequency ω with mode index m, n, are defined in such a way that

$$\vec{E}_{m,n}^{(\omega)}(\vec{r}, t) = \frac{1}{2} \left\{ A_{m,n}^{(\omega)}(z) \vec{\mathcal{E}}_{m,n}^{(\omega)}(x, y) e^{i(\omega t - \beta_{m,n} z)} + c.c. \right\}$$

holds for a waveguide mode propagating in z-direction. $\vec{\mathcal{E}}_{m,n}^{(\omega)}(x, y)$ describes the normalized electric field distribution in the waveguide cross-section.

It is sufficient to consider the normalized field amplitude $A_{m,n}^{(\omega)}$ and z-dependence of a mode to describe the interaction in a homogeneous waveguide with a refractive index cross-section $n(x, y)$. A generalized index i is now introduced to describe the interaction between three well-defined modes at frequencies ω_i with $i = 1, 2, 3$. Then, the CME can be written as follows [23]:

$$\frac{\partial A_1}{\partial z} = -i\kappa_1 \nu^* A_2^* A_3 \exp(-i\Delta\beta z) - \frac{\alpha_1}{2} A_1$$
$$\frac{\partial A_2}{\partial z} = -i\kappa_2 \nu^* A_1^* A_3 \exp(-i\Delta\beta z) - \frac{\alpha_2}{2} A_2 \quad (1.11)$$
$$\frac{\partial A_3}{\partial z} = -i\kappa_3 \nu A_1 A_2 \exp(i\Delta\beta z) - \frac{\alpha_3}{2} A_3.$$

Power in the respective modes is given by $\int\int I_{m,n}^{(\omega)}(\vec{r}, t)\, dxdy = P_i$. Waveguide propagation losses are considered in the damping coefficient α_i, and the three equations are coupled by the coupling term $\kappa_i = 2\pi d_{ijk}\sqrt{2\left(n_1 n_2 n_3 c\epsilon_0 \lambda_i^2\right)^{-1}}$, with the effective indices n_i. The overlap integral is given by

$$\nu = \int_0^\infty \int_{-\infty}^\infty \mathcal{E}_1(x, y)\mathcal{E}_2(x, y)\mathcal{E}_3(x, y)dxdy$$

which accounts for the effective interaction area determined by the electric field distributions of the interacting modes. d_{ijk} is an element of the nonlinear tensor which is coupling the three considered waves i, j, k of arbitrary polarization. The complex conjugates are marked with an asterisk. The phase-mismatch term is given by $\Delta\beta = \beta_3 - \beta_2 - \beta_1$. The system of coupled equations may be solved analytically, if we assume identical (power) losses α_i for each wave, as well as a non-depleting pump, i.e. pump losses are given by propagation losses only and not by conversion. For a detailed treatment, see e.g. [9, 15].

For now it shall be sufficient to consider a phase-matched interaction with zero losses, where we assume $\Delta\beta$ and α_i to be zero, as well as an effective nonlinear coefficient d_{eff} of a given value (see section 1.1.3 for more information). In this case, for SFG we get

$$A_2(z) = A_2^0 \quad \text{(pump)}$$
$$A_1(z) = A_1^0 \cos\left(\sqrt{\kappa_1\kappa_3}|\nu|\,|A_2^0|\,z\right) \quad \text{(signal)} \quad (1.12)$$
$$A_3(z) = -i\sqrt{\frac{\lambda_1}{\lambda_3}}A_1^0 \sin\left(\sqrt{\kappa_1\kappa_3}|\nu|\,|A_2^0|\,z\right) \quad \text{(sum frequency)}$$

Note that the input phases of pump and signal are preserved during frequency conversion and add up to an additional phase of the sum frequency [24]. From equation (1.12) directly follows

$$P_2(z) = P_2^0 \quad \text{(pump)}$$
$$P_1(z) = P_1^0 \cos^2\left(\sqrt{\kappa_1\kappa_3 P_2^0}\,|\nu|\,z\right) \quad \text{(signal)} \quad (1.13)$$
$$P_3(z) = \frac{\lambda_1}{\lambda_3}P_1^0 \sin^2\left(\sqrt{\kappa_1\kappa_3 P_2^0}\,|\nu|\,z\right) \quad \text{(sum frequency)}$$

for the power evolution. In the case of DFG, we have

$$A_3(z) = A_3^0 \quad \text{(pump)}$$
$$A_2(z) = A_2^0 \cosh(\sqrt{\kappa_1\kappa_2}|\nu|\,|A_3^0|)z \quad \text{(signal)} \quad (1.14)$$
$$A_1(z) = -i\sqrt{\frac{\lambda_2}{\lambda_1}}A_2^0 \sinh(\sqrt{\kappa_1\kappa_2}|\nu|\,|A_3^0|)z \quad \text{(idler)}$$

and, for the power evolution,

$$P_3(z) = P_3^0 \quad \text{(pump)}$$
$$P_2(z) = P_2^0 \cosh^2(\sqrt{k_1 k_2 P_3^0}|\nu|)z \quad \text{(signal)} \quad (1.15)$$
$$P_1(z) = \frac{\lambda_2}{\lambda_1}P_2^0 \sinh^2(\sqrt{\kappa_1\kappa_2 P_3^0}|\nu|)z \quad \text{(idler)}$$

In order to estimate conversion efficiencies, we can now insert some realistic parameters into equations (1.13) and (1.15). Assume that an MIR wavelength (here: 3.4 μm) should be converted to the NIR to be detected with an InGaAs detector. In the DFG case, a pump at 1064 nm wavelength is chosen to generate the DF at 1550 nm; in the SFG case, the pump at 1550 nm wavelength is chosen to generate the SF at 1064 nm. If we assume feasible power levels for pump and signal, and take calculated values for the overlap integral ν ($\approx$ 72000 m^{-1} [12]) and coupling coefficients κ_i, we get the curves displayed in figure 1.3 for the converted power.

From these curves we can see that, in the regime of short interaction lengths and low pump power, the conversion efficiency, defined by $\eta = \frac{P_{i/sf}(L)}{P_p(0)P_s(0)}$, is comparable for both

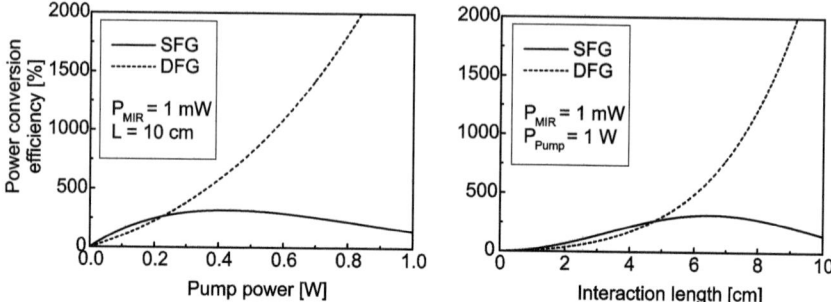

Figure 1.3.: Dependence of SFG/DFG power conversion efficiencies as function of Length (left) and pump power (right), using a simple analytical model of a Ti:PPLN waveguide.

processes (it is, in fact, larger for the SFG-process, mainly because of the fact that higher energy photons are generated than in the DFG case). In this case, conversion efficiency scales nearly linearly with pump power, and it may be estimated that for small z, P_p and P_s

$$\frac{d^2 P_{SF}}{dP_p dP_S} = \frac{2}{n_p n_s n_{SF}} \sqrt{\frac{\mu_0}{\epsilon_0}} \left(\frac{2\pi d_{eff} \nu}{\lambda_{SF}} L\right)^2 = \eta_{SF}[1/W]$$

$$\frac{d^2 P_{DF}}{dP_p dP_S} = \frac{2}{n_p n_s n_{DF}} \sqrt{\frac{\mu_0}{\epsilon_0}} \left(\frac{2\pi d_{eff} \nu}{\lambda_{DF}} L\right)^2 = \eta_{DF}[1/W]$$

Plugging realistic values, we get

$$\eta_{SF} = 1335 \frac{\%}{W} \propto \frac{1}{\lambda_{SF}^2} \propto \omega_{SF}^2$$

$$\eta_{DF} = 629 \frac{\%}{W} \propto \frac{1}{\lambda_{DF}^2} \propto \omega_{DF}^2$$

At higher pump powers, respectively interaction lengths, however, we observe an exponential slope due to $\sinh^2$-type behavior in the case of DFG, while conversion saturates in the case of SFG due to $\sin^2$-type behavior. The reason for this is so-called parametric amplification in the case of DFG: due to the generation of an addition signal photon during the decay of each pump photon, the nonlinear process is effectively enhanced. In reality, the pump will deplete however, due to the nonlinear process as well as propagation losses, so the effect is somewhat limited. Pump depletion can be accurately modeled by solving equations (1.11) numerically.

In the SFG-case, the process is reversed beyond unit conversion. If sufficient pump power is available in a given waveguide geometry, all the incident photons are up-converted to the sum-frequency. A further increase of pump power will lead to the re-generation of signal photons together with the pump - the reason is given by the fact that the direction of energy transfer between modes is dictated by a phase relationship. The phase is determined automatically if one of the interacting waves has zero amplitude, according to equations 1.12 and 1.14.

From these basic considerations, assuming similar detector performance at the respective SF and DF wavelengths, as well as equivalent powers at the respective pump wavelengths, we can derive the following: If sufficient pump power and sufficiently long interaction lengths (compromising, however, phase-matching bandwidth) are available to reach the regime of notable parametric amplification, DFG is the interaction of choice. However, SFG has a slight advantage in power conversion efficiency at lower power levels, due to conversion to higher energy photons. In reality, available instrumentation (detectors, pump sources), as well as spectral filtering need to be taken into account (see next chapter for a detailed discussion).

There is, however, another catch. In the case of the DFG-interaction, amplified parametric fluorescence leads to the generation of photons at the up-conversion wavelength. The consequences for the up-conversion detector are discussed in the chapter 3.

1.1.3. Quasi Phase-Matching

We have taken $\Delta\beta$ as zero above, assuming perfect phasematching. However, due to material dispersion, there is generally a phase mismatch which needs to be compensated. Birefringence can be used for this purpose, but quasi-phasematching (QPM) by periodic poling is generally favored because it enables access to the highest nonlinear coefficient $d_{33} \approx 20$ pm/V of $LiNbO_3$ [9] and enables nearly arbitrary wavelength combinations. It is even possible to use counter-propagating interactions, which was demonstrated in 2007 by Canalias et al. in the form of a mirrorless OPO [25]. The need for these phase-matching techniques leads to a limited bandwidth of the nonlinear process, which will be discussed in later chapters (a detailed discussion is also found in [12]).

In Ti-indiffused waveguides in Z-cut, Y-propagation $LiNbO_3$, the d_{33}-coefficient leads to nonlinear optic interactions between TM-polarized waveguide modes. Recalling equation (1.11), we can develop the discrete modulation due to periodic poling of the substrate by

$$d_{eff}(z) = d_{ijk} \sum_{m=-\infty}^{\infty} G_m \exp(-iK_m z) \qquad (1.16)$$

with the wave vector K_m of the periodic grating structure and Fourier coefficients $G_m = \frac{2}{m\pi} \sin(m\pi)$ (for a fixed periodicity and duty cycle). If we plug equation (1.16) into equation (1.11) and neglect higher order coefficients, d_{ijk} is replaced by $d_{eff} = \frac{2}{\pi} d_{ijk} \exp(-i\frac{2\pi}{\Lambda} z)$ and so the phase mismatch term in equation (1.11) becomes

$$\Delta\beta = \Delta\beta_3 - \Delta\beta_2 - \Delta\beta_1 - \frac{2\pi}{\Lambda} \qquad (1.17)$$

which should become zero, with the poling periodicity Λ. Λ must be determined according to the desired nonlinear interaction. Solving equation (1.11) using this approximation

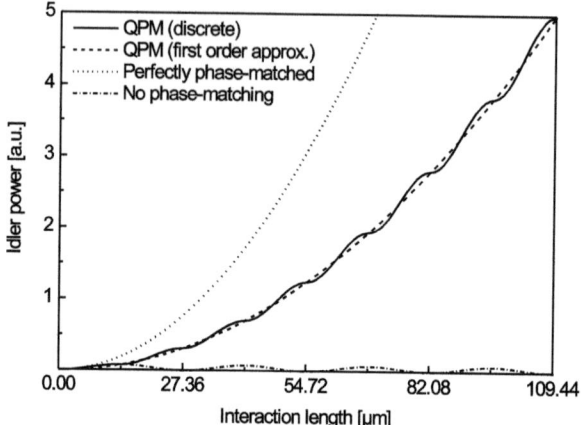

Figure 1.4.: Nonlinear-optic generation of radiation in an optical waveguide over short interaction lengths. In case of a one-pass nonlinear optical interaction with interaction lengths in the range of centimeters, the first-order QPM approximation describes the process exceptionally well.

describes reality sufficiently well for interaction lengths much longer than Λ; however, the higher order coefficients in equation 1.16 lead to an additional modulation of the power conversion which is evident for very short interaction lengths. This effect can be seen in figure (1.4). The calculated domain periodicity is $\Lambda = 27.36$ μm. After each half period of Λ, the sign of the nonlinear coefficient is changed, corresponding to a phase-shift of π according to equation 1.17. The periodic change of the sign of the nonlinear coefficient enables a continuous power conversion. Without phase-matching, power is exchanged periodically between the three waves without efficient frequency generation.

A curve for perfect phase-matching without QPM, assuming the same nonlinear coefficient, is also given for comparison. It is possible to achieve e.g. birefringent phase-matching without QPM, however the nonlinear coefficient is considerably lower when the interaction of orthogonally polarized waves is considered.

1.2. Waveguide Sample Fabrication

After a $LiNbO_3$ substrate has been prepared, the sample fabrication consists of two main fabrication steps: In the case of Ti-waveguides, waveguides are indiffused first, followed by a period poling process. Afterwards, some finalizing steps, like end-face polishing, are usually also necessary. The fabrication is described in the following in some more detail [26]:

1.2.1. Waveguide Fabrication

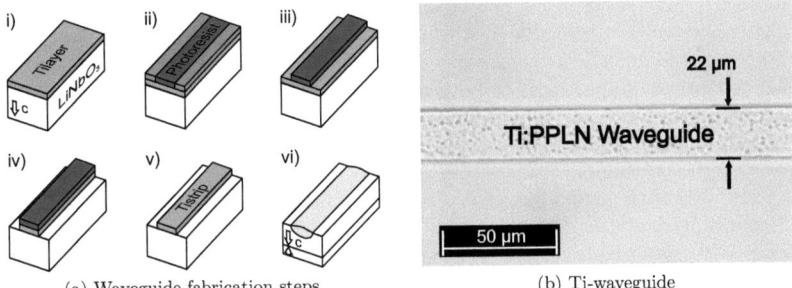

(a) Waveguide fabrication steps (b) Ti-waveguide

Figure 1.5.: (a) Waveguide fabrication steps: i) Ti-deposition; ii),iii) photoresist structuring; iv),v) Ti-strip formation; vi) Ti-indiffusion. (b) Top-view micrograph of a Ti-indiffused MIR waveguide.

In order to fabricate waveguides in $LiNbO_3$, Titanium stripes may be indiffused into a crystal substrate in order to form strip waveguides. The exact fabrication parameters are usually chosen in such a way that, if possible, a waveguide which is single-mode at the design wavelength(s) is obtained. A number of steps need to be taken before the Titanium is indiffused at elevated temperatures:

The first step is to cut a ca. 90 x 12 x 0.5 mm^3 sample substrate from a 4" $LiNbO_3$ wafer. The sample is coated with a Ti-layer by E-beam evaporation (figure 1.5a (i)); for MIR waveguides, the layer thickness is about 170 nm. In order to create the waveguide structure, a positive photoresist layer (type AZ6615) is then spin-coated onto the sample surface and baked at 90°C for 30 minutes (figure 1.5a (ii)). Using a chromium photomask as the template, the photoresist is then structured by photo-lithography. The structure consists of eight identical waveguide groups, including a tapered region for fiber coupling, as well as labels and alignment marks. The photoresist is developed using type AZ400K developer (figure 1.5a (iii)). Afterwards, the exposed Ti is wet-etched using EDTA (figure 1.5a (iv)) and the remaining photoresist is removed in an acetone bath. (figure 1.5a (v)). In order to indiffuse the Ti into the substrate to create the dielectric waveguides, the sample is placed into a programmable oven for 31 hours at 1060°C, including warm up and cool down times (figure 1.5a (vi)).

The indiffused Ti induces a refractive index profile in the substrate, enabling wave guidance. Some patchy structures form on top of the waveguides during the indiffusion process (figure 1.5b), and the resulting surface roughness contributes to waveguide propagation losses. Still, losses have been found to be about 0.05 dB/cm for Ti-indiffused, MIR waveguides at 3.4 μm, which is still exceptional [12]. Also, at the bottom of the substrate, the c-axis (corresponding to the crystallographic Z-axis) is spontaneously flipped in a roughly 10 μm thick layer due to the pyroelectric effect during indiffusion [9].

1.2.2. Periodic Poling

After the waveguides have been fabricated, the substrate needs to be periodically poled in order to later on facilitate quasi-phasematching in the waveguides. The poling periodicity is calculated beforehand, which is done using semi-empirical models of material dispersion and numerical calculations of waveguide mode dispersion (see section 1.1.1). Due to fabrication tolerances and other incalculable deviations between experiment and theory, a set of varying poling periodicities is usually used on a single sample. This allows to finally choose the waveguide with the best nonlinear properties at a desired wavelength and suitable temperature.

The fabrication method of choice is electric field assisted periodic poling. As the growth of ferroelectric domains begins at the +Z side of the crystal, the bottom layer needs to be removed by careful polishing (figure 1.6a (i),(ii)). Afterwards, the polarity of the whole sample needs to be reversed. For this purpose, it is placed in between two rubber O-rings within a specially designed poling holder. The area within the O-rings is then filled with a lithium chloride liquid electrode (figure 1.6a (iii)) and an electric field higher than the coercive field strength of $LiNbO_3$ (ca. 21 kV/mm [27]) is applied to 'flip' the c-axis (figure 1.6a (iv)). During this step, the current must be recorded to estimate the necessary charge for the actual periodic poling step, which is reduced due to the poling mask when compared to the 'flipping'-charge.

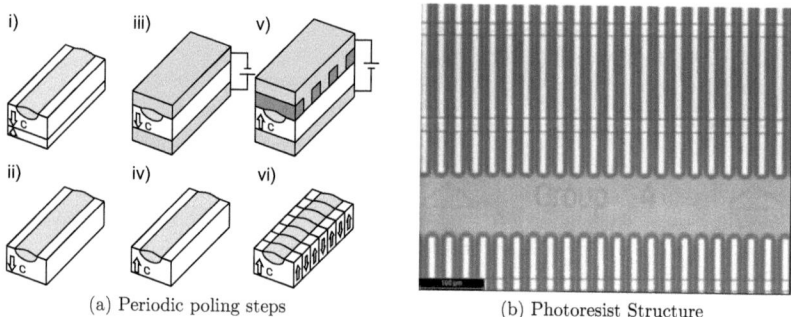

(a) Periodic poling steps   (b) Photoresist Structure

Figure 1.6.: (a) Periodic poling steps: i),ii) removal of spontaneous polarization layer; iii),iv) c-axis reversal; v),vi) periodic poling using a lithographic grating structure. (b) Top-view micrograph of Ti:PPLN sample before periodic poling. Horizontal lines are the waveguides, below the photoresist layer.

Again, a photo-lithographic structure is needed in order to create the poling structure. A photoresist (type AZ4533) layer is again spun onto the sample surface, this time with a slightly thicker resist used to effectively shield the electric field during the poling process. The resist is structured by photolithography (figure 1.6b). The mask layout is organized in 6 poling groups with periodicities from $\Lambda_1 = 26.5$ μm to $\Lambda_6 = 27.25$ μm in 0.15-μm steps (see appendix A for details). The mask should be carefully aligned to the waveguide structures, which were previously created. After baking the photoresist at 90°C for 30 minutes, the photoresist is developed (figure 1.6a (v)) and the poling procedure is repeated, this time generating the periodic poling structure (figure 1.6a (vi)).

Waveguide fabrication is now completed and the sample end faces can be polished. Afterwards, additional dielectric anti-reflection coatings, as well as pump reflection mirrors, may be applied as needed, using evaporation deposition techniques.

1.3. Summary and Discussion

The indiffusion of Ti-stripes into a LiNbO$_3$ substrate is a mature processing technology to realize low-loss optical waveguides. Waveguide propagation losses below 0.1 dB/cm can be readily achieved. Together with electric-field assisted periodic poling, to generate quasi-phasematched Ti:PPLN waveguide samples, efficient non-linear devices for MIR applications can be realized. Theoretically, conversion efficiencies of the order of several hundred %/Watt are feasible for a sample of several centimeters length. The bandwidth of the nonlinear process is of the order of nanometers, as will be seen in later chapters.

Note that there are some intrinsic effects which limit performance - as well as room temperature operation - of these devices. This includes:

- The large waveguide cross-section of an MIR waveguide, which is multimode at NIR wavelengths [12]. Due to the nature of nonlinear interactions in a waveguide, the selective excitation of fundamental modes of pump and/or signal radiation is imperative, and may be realized using a carefully designed taper at the fiber-connected end of the waveguide. Corresponding taper-sections were implemented in the mask design used for the lithographic waveguide structuring. Also, careful adjustment of coupling optics is just as crucial.

- The waveguide index profile is inherently asymmetric, due to the surface-bound indiffusion process of Ti into the substrate. This leads to a wavelength-dependent position of (fundamental) waveguide modes, in addition to differing mode sizes. The mode distributions determine the overlap integral, which in turn dictates the efficiency of a waveguide-based nonlinear process. Well-centered modes would be preferable, due to an improved overlap integral.

- Ti-doped LiNbO$_3$ shows pronounced photorefractive properties [28], i.e. an optical absorption-induced variation of the refractive index distribution in a waveguide is observed. Also known as optical damage, this has a strong effect on non-linear interactions in waveguides, mainly because the phase-matching term $\Delta\beta$ depends on refractive index. Photorefractivity can be mitigated by heating of the optical sample to elevated temperatures. The Ti:LiNbO$_3$ waveguide samples were operated at temperatures above 150°C to achieve stable output power and repeatable tuning characteristics.

There are promising developments to circumvent such difficulties. Zn-doped $LiNbO_3$ has been used for example to fabricate ridge waveguides which are optical damage resistant and offer high MIR power output even at room temperature [29]. Also, the use for example of ridge waveguides, or in the future possibly even of periodically-poled photonic wires [30], can improve mode confinement as well as mode overlap for an optimized nonlinear optical waveguide interaction. For these techniques to be competitive, however, fabrication techniques still need to be improved to realize low-loss waveguides of consistent quality.

2. Mid-infrared Source based on Difference-Frequency Generation

Nonlinear wavelength conversion of radiation is an attractive means to reach certain wavelengths in the MIR-range, as has been previously demonstrated by several research groups, see e.g. [10, 29, 31]. In fact, DFG-MIR-sources with wavelengths around 3.4 μm have matured to commercial status and are currently being marketed by a small number of companies, using bulk crystals [32] and even waveguides [33].

In the past, the MIR range above 3-μm wavelength was covered mainly by lead-salt-compound lasers - these are, however, limited in tuning range by mode hops, and limited in output power by low thermal conductance of the materials [7, 34]. QCL's represent a more recent and advanced development in the area of MIR laser sources. At the time of writing, QCL's are commercially available and even wavelength tunable external cavity QCL's, using the QCL structure as gain medium, have reached commercial status [35, 36]. The QCL was demonstrated by Feist et al. in 1994 [37], who later also realized the first tunable external cavity QCL [38].

However, commercially available QCL's cover the wavelength range mainly above 4.5 μm. Emission around 3.4 μm has been demonstrated in the laboratory before [39], albeit cw-operation (cw = continuous wave) was limited to cryogenic temperatures.

DFG sources using frequency conversion in single-mode waveguides exhibit a naturally good mode quality, flexible tunability, and power scalability. Using DFG, nearly arbitrary wavelengths may be reached by mixing suitable wavelengths. Output powers and spectral characteristics are mainly determined by the available pump and signal sources, which can have excellent properties with respect to output power and spectral purity in the NIR range themselves. Several tens of milliwatts of generated MIR power have been achieved by down-conversion to 3.4 μm using DFG in diced ridge waveguides [29].

Wavelength tuning is facilitated by tuning of either or both of pump and signal sources. If only one of either is tuned, the phase-matching bandwidth limits the useful instantaneous tuning range of a DFG-MIR source. However, synchronized tuning of both temperature and signal wavelength was demonstrated in bulk optic crystals for wide-range wavelength scans [31]. In this case, a PPLN crystal was heated to elevated temperatures and then cooled down by simply switching off the heating current. The temporal phase-matching characteristics of the crystal was then determined in rigorous characterization of the cooling process; subsequently, the signal wavelength could be adjusted as a function of time in order to retain phase-matching.

In addition to the attractive modal quality and power scalability of a DFG source, the non-linear wavelength conversion retains not just amplitude information of a down-converted signal, but also the phase-information; a feature which is especially useful for phase-encoded data transmission.

In the previous chapter, the theoretical background and fabrication steps for wavelength conversion devices based on nonlinear optical interactions in waveguides were discussed. In this chapter, an MIR source using DFG is described and experimental results, including conversion efficiency and tunability, are presented. In this case, 1550-nm signal radiation is mixed with a 1064-nm pump in order to generate 3.4-μm radiation.

2.1. Theoretical Modelling

In order to estimate the performance of a DFG-based MIR radiation source, modeling according to section 1.1 was performed. The nonlinear interaction was modeled for an 18-μm wide, Ti-indiffused waveguide of 90 mm interaction length. First, the overlap integral ν was determined by calculating the electric field distributions of the three interacting waveguide modes using FOCUS [16]. The nonlinear coefficient d_{33} was calculated using Miller's Δ. In literature, the values of nonlinear coefficients are mainly given for second-harmonic generation in the visible to near-infrared wavelength range, so Miller's Δ was used in order to calculate the nonlinear coefficient for the respective SFG and DFG interactions of interest here [40]. Miller's Δ assumes a singly resonant harmonic oscillator model, in which case it has a constant value:

$$\Delta_{ijk} = \frac{d_{ijk}(-\omega_3; \omega_1, \omega_2)}{[n_i^2(\omega_3) - 1][n_j^2(\omega_1) - 1][n_k^2(\omega_2) - 1]}.$$

At the given wavelength combination of 3400 nm, 1550 nm, and 1064 nm, and assuming a coefficient $d_{33} = 25.2$ pm/V for SHG at 1064 nm, we get $d_{33} = 17.8$ pm/V.

The following calculations were performed by plugging given values into equation (1.11) and assuming quasi-phasematched interactions. Waveguide losses of 0.05 dB/cm at 3.4 μm, 0.1 dB/cm at 1550 nm and 0.2 dB/cm at 1064 nm were assumed [12]. The first value was measured according to the Fabry-Pérot fringe contrast method [41] with a reference PPLN sample; however the two values at the shorter wavelengths were estimated because the contrast measured is only applicable to mono-modal waveguides. The losses are assumed to be higher because scattering losses increase with wavelength.

The results of the calculation are shown in figure 2.1. The waveguide-internal idler power at the end of the 90-mm interaction length is given as a function of pump and signal powers at the waveguide input. In contrast to the simplified considerations previously discussed and presented (see figure 1.3), the calculations were now performed numerically and consider pump depletion as well as realistic waveguide losses. The pump depletes as pump photons decay to signal and idler photons - as a result, the DFG-process is limited by available pump power. At a given initial pump power, the maximum idler which can be extracted from the Ti:PPLN waveguide is reached when the pump is completely depleted by mixing with a sufficiently strong signal wave. If

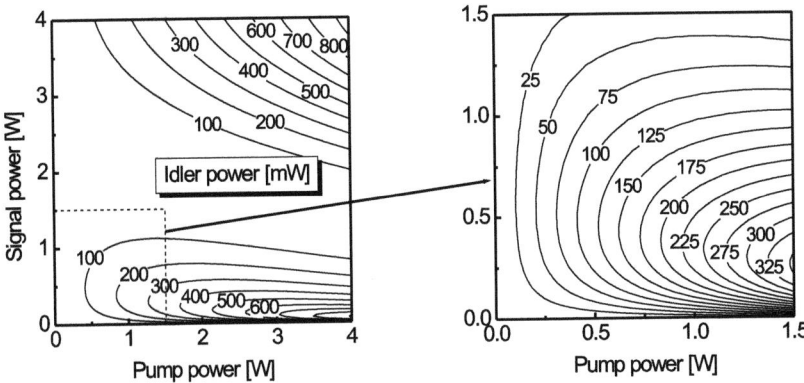

Figure 2.1.: Calculated power dependence of the generated idler radiation on pump and signal powers in a 90-mm waveguide, considering propagation losses and pump depletion. Left: up to 4 Watts of pump/signal radiation, right: detailed view of up to 1.5 Watts of in-coupled pump/signal radiation.

input signal power is increased any further, the nonlinear process is reversed within the interaction length, so to speak, and pump-photons are 're-generated' by SFG between signal and idler photons.

This may be explained by the following consideration: Phase-matching is possible for both SFG and DFG, as indeed both processes are described by the same equations. If one of the waves is depleted, its phase becomes undefined - corresponding to the initial problem, which is usually solved for in nonlinear optical interactions. Only the phase-relation between the three interacting waves determines whether SFG or DFG takes place (cp. equations 1.12, 1.14 - the generated wave lags behind with an additional phase factor of $-i$ in each case).

The maximum permitted power is given by the destruction threshold of the waveguide. The threshold for bulk $LiNbO_3$ is in the range of 100 MW/cm^2 [42], so assuming a modal cross section of 4 μm^2 at the pump wavelength of 1064 nm, this corresponds to a maximum pump power of about 4 Watts. In case of the waveguide geometry discussed, the maximum idler power is then expected from theory with just 60 milliwatts of signal power at 4 Watts of pump power, generating more than 900 milliwatts of idler. At lower pump power levels, the signal power needed to deplete the pump increases slightly due to reduced parametric signal amplification. Higher output power levels are in theory possible within the 'second maximum' along the signal-power axis, however the highest conversion efficiency is obviously reached within the fundamental conversion peak at low signal power levels. Such 'extreme' situations did not occur experimentally due to the moderate power levels.

A number of additional factors restrict the maximum permitted power below the assumed damage threshold of 4 Watts, being

- injection efficiency and damage resilience of the waveguide interface,
- photo-refractive damage, which becomes significant at milliwatt-range powers at 1064 nm and room temperature operation, which seriously degrades performance due to changes of the phase-matching condition within the waveguide [28],
- power allowance of all the other optical components involved.

In this work, waveguide-internal power levels were limited to the order of 100 milliwatts of pump and signal, meaning that pump depletion and other aforementioned effects did not pose serious difficulties. MIR output-powers above 10 milliwatts at 3.8-μm wavelength were demonstrated experimentally using fiber-coupled lasers as pump and signal sources [43]. Output power was limited by fiber-optic components, which were used as well to combine pump and signal radiation, as to couple radiation to the waveguide samples. In addition, insertion losses of the fiber-optic components as well as coupling losses to the waveguide reduced the power available in the waveguide.

Asobe et al. have demonstrated even higher output powers of a MIR-DFG source, using a Zinc-doped and thus damage resistant $LiNbO_3$ ridge waveguide fabricated by direct bonding and dicing technique [29]. They could generate 65 milliwatts of idler at 3.4-μm wavelength, with 444 milliwatts of injected pump at 1064-nm wavelength, and 558 milliwatts of injected signal at 1550 nm.

2.2. Experimental Setup

An MIR source using DFG requires the following main components, shown in figure 2.2a: Firstly, the nonlinear optical crystal, and secondly, pump and signal laser sources. A tunable external cavity laser (ECL) with a tuning range from 1500 to 1600 nm in 1 pm steps and output powers in the milliwatt range was used as signal source. As for the pump source, a fiber Bragg grating (FBG) stabilized laser diode (LD) operating at 1064 nm, with an external fiber optic isolator to suppress feedback-related output fluctuations, was used as pump for 3.4-μm wavelength generation. The utilizable output power (behind isolator and polarization control) was 32 milliwatts. Signal power may easily be amplified to several Watts, using an erbium doped fiber amplifier (EDFA) behind the ECL. However, signal power was restricted to more conservative values most of the time in order to protect fiber optic components. Figure 2.2b shows a Ti:PPLN sample housed in a Cu-oven in an isolating Teflon™ enclosure. The fiber coupling stage is also shown in the figure.

For a brief period of time in this work a high-power pump at 1064-nm was available (Klastech Senza [44]). In case of communication experiments discussed in chapter 5, a high-power, 1100-nm source was used to generate and re-convert 3.8-μm radiation.

Light can either be coupled into (or out of) the waveguide using lenses, or directly from a fiber by butt-coupling. Coupling efficiency is determined by the modal overlap

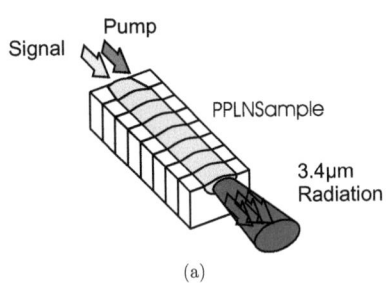 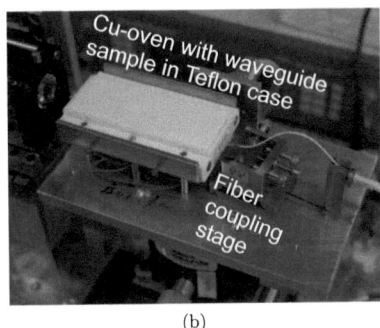

Figure 2.2.: (a) Illustration of a DFG-MIR source. (b) Photo of a Ti:PPLN sample housed in a Cu-oven, isolated by a Teflon^{TM} enclosure. The fiber coupling stage of the DFG-MIR module is also visible.

between a waveguide mode, and the field distribution of the impinging radiation at the respective wavelength. In order to selectively excite the fundamental modes at the 'NIR'-input, a linear taper section of 3 mm length is used, where the waveguide width increases from 4 μm to full width (see section 1.2.1, figure 1.5a).

In the case of frequency mixing, one main inconvenience when using lenses to couple light into the waveguide is chromatic dispersion of the lens material. This has been illustrated in figure 2.3: here, two lenses are shown; one for beam collimation, and one for coupling to the waveguide. In the Gaussian beam picture, the differing refractive indices of the first lens will lead to beam waists at two different positions. This in turn causes a slight shift of the focal distance behind the coupling lens, which can easily surpass the Rayleigh length (see appendix B for a short discussion). Even more severe are lateral shifts of any optical component, which will lead to harmful deviations of the two beams within the focal plane.

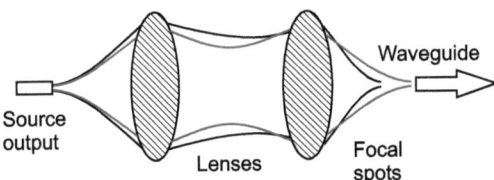

Figure 2.3.: Illustration of lens dispersion effects when coupling different wavelengths to a waveguide.

In the optimum case using dispersive lenses, the beams coming from both lasers would first need to be collimated individually, using lenses operating at their design wavelengths. The two beams could then be combined using a beam combiner (realized using e.g. using a specially designed dielectric layer system on a transparent substrate) in order to couple both to the waveguide. The overlap of the free-space propagating modes with the waveguide modes could be optimized individually for the two wavelengths.

It was found that the nonlinear efficiency was several times higher using an optical fiber for coupling, than using a simple lens arrangement as shown in figure 2.3, due to the reasons discussed above. Also, the use of fiber optic components together with waveguides has a number of additional advantages:

- complexity is minimized because functionalities can be combined,
- the need for bulky components is reduced,
- mechanic stability requirements may be relaxed due to miniaturization,
- compatibility to fiber-optic devices is given.

In the fiber optic setup, a WDM coupler is used to combine the two incident wavelengths, and a glass ferrule mounted on a miniaturized translation stage is used to couple the light to the waveguides (figure 2.4).

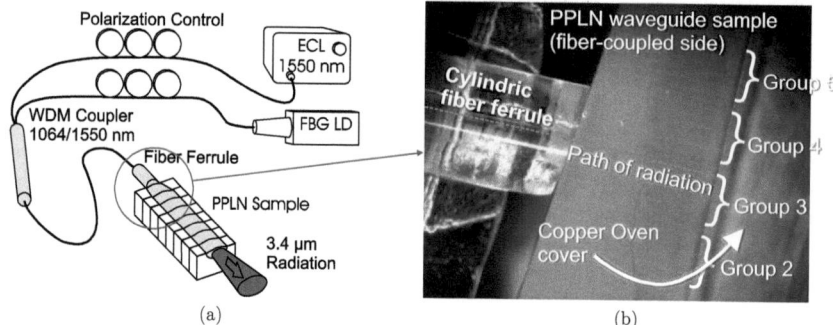

Figure 2.4.: (a) DFG setup using fiber-coupling to the waveguide. Take note of the exaggerated scale of the waveguide. (b) Microscopic view of the fiber ferrule used for fiber-coupling. Several waveguide groups, each with 6 waveguides, are visible. The path of the radiation coming from the fiber, which is coupled to an individual waveguide (waveguide 6 of group 3, in this case) is sketched by arrows. The bottom side of the sample has a rough polish, making the sample appear whitish.

Both fiber ferrule and waveguide sample are polished at respective angles in order to suppress back reflections and interference fringes both inside and outside of the sample. The sample end face is polished at a 5.8° angle in order to match the emergence angle from the ferrule, which is polished at an 8° angle. Also, a broad-band anti-reflection (AR) coating was applied to the fiber-coupled side of the waveguide in order to reduce fringing effects due to Fabry-Perot cavity resonances (see figure 2.5). The sample itself is kept in a copper oven which can be heated to above 200°C in order to alleviate photo-refractive damage due to the high optical field strengths in the waveguide.

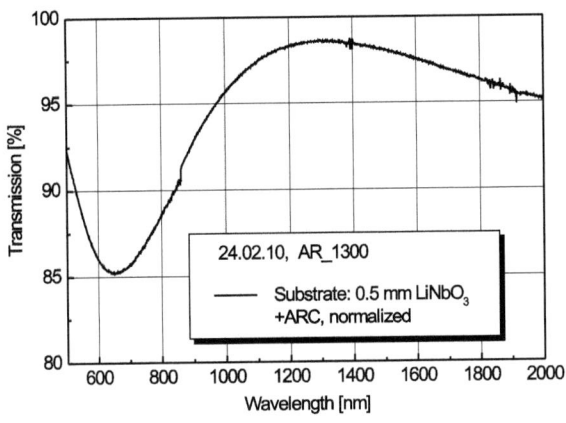

Figure 2.5.: Transmission properties of the anti-reflection coating of the fiber-coupled endface (measured on a witness sample). The coating is centered at 1300 nm, offering satisfactory performance at both 1064 and 1550 nm.

The nonlinear interaction in the waveguide will generate idler radiation efficiently only when the phase-matching condition is fulfilled. Phase-matching can be achieved by either tuning sample temperature, or by scanning the signal (or pump) wavelength. Both tuning mechanisms lead to changes in the differential phase of the interacting waves. The reason is the temperature- and wavelength-dependence of the refractive index as well thermal expansion effects. Tuning of either, temperature or signal (pump) wavelength, is in principle suitable to record the phase-matching characteristics of a device, if the pump (signal) wavelength is fixed. When both, temperature and wavelength are tuned in unison, the phase-matching condition can be maintained over a long wavelength range. This procedure has been used in bulk-optic DFG crystals for broadband wavelength sweeps in the mid-infrared to detect constituents of surgical smoke by Gianella et al. [31].

Temperature tuning during operation is more or less unsuitable when a waveguide-based frequency conversion device is used, due to thermal expansion of the material which will alter the coupling efficiency. This effect would be alleviated by pigtailing the fiber to the waveguide sample, which was however not (yet) done due to the high operating temperatures of the waveguide sample. Thus, the usual procedure to find the

phase-matching condition is to
- stabilize the sample to a desired temperature,
- optimize fiber coupling to the waveguide at this specific temperature,
- scan the signal wavelength while monitoring the output using an MIR detector.

The MIR beam emerging from the waveguide end face may be collimated by a suitable lens. For this purpose, a ZnSe best-form lens with a focal length of 15 mm is used in the experiments. Pump and signal radiation emerging from the waveguide must be blocked; an AR-coated Germanium filter, which is transparent in the MIR, is used to fulfill this purpose. An MCT detector is used for detection.

2.3. Sample Characterization & Discussion of Results

The characterization of the Ti:PPLN waveguide samples was restricted to the following:
- investigation of MIR waveguide mode profiles,
- determination of external and internal conversion efficiencies,
- investigation of phase-matching properties / wavelength tuning characteristics, and
- power stability measurements.

The results are presented in the following:

2.3.1. Mid-infrared Mode Distributions

The photo-mask used for waveguide fabrication contains eight waveguide groups with six waveguides each. The waveguide groups have an identical structure, with two waveguides each of widths 18 μm, 20 μm and 22 μm (see appendix A). The former two exhibit monomode behavior at 3.4-μm wavelength (QTM$_{00}$-mode), while the latter also supports the first order QTM$_{10}$ mode. This was verified using a Stirling-cooled MIR camera [45], with a 320 × 256 pixel InSb detector array with 30-μm pixel pitch (see figure 2.6). Radiation from the 3.4-μm HeNe-laser was coupled into and out of the waveguide using f = 15 mm, ZnSe bestform lenses with a calculated resolving power of 4.4 μm.

In figure 2.7, a measured mode profile from an MIR waveguide is compared to a calculated intensity profile using FEM (Finite Element Method) in FOCUS [16]. In the calculated profile, a pronounced asymmetry is apparent near the waveguide surface (given by the horizontal line at y = 0 μm), due to the high index contrast to air. Towards the substrate, the index gradient is much smaller due to the indiffused Titanium, so the mode spreads out into the substrate. In the measured profile, the effect is not visible which is a strong indication that the ZnSe lens resolution is insufficient to fully resolve the features of the near-field being emitted from the waveguide. This has some implications on the coupling efficiency of an MIR-beam *to* such a waveguide, which is of importance for the free-space transmission lines discussed in later chapters.

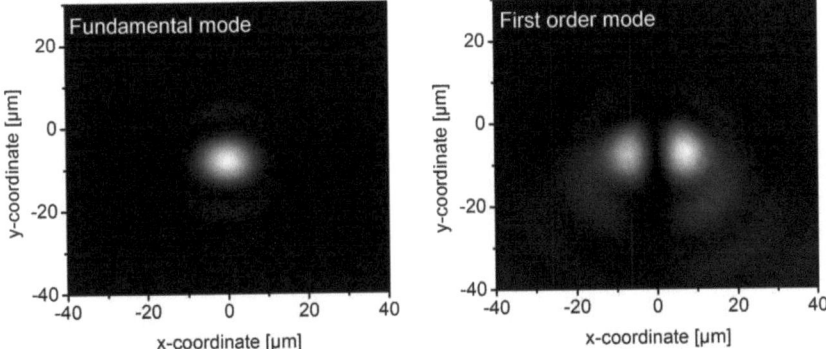

Figure 2.6.: Measured intensity distributions of fundamental mode QTM_{00} (left) and first order, QTM_{10} mode (right) in a 22-μm waveguide. Radiation coming from a HeNe laser emitting at 3.39-μm wavelength was selectively coupled to the respective waveguide mode by mechanical adjustment of the input lens. The mode was imaged onto the MIR camera using an AR-coated, f =15 mm ZnSe lens.

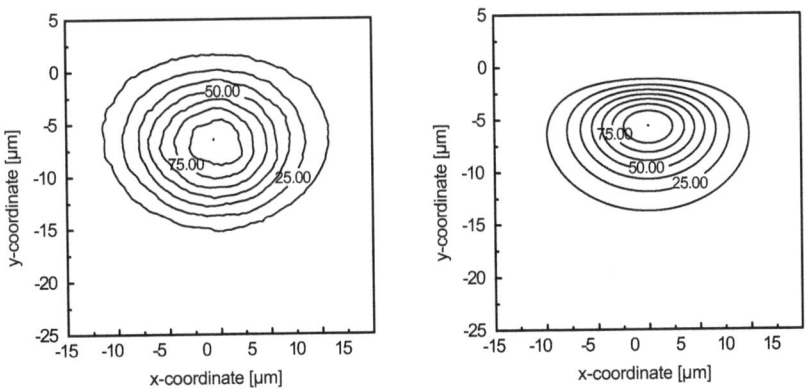

Figure 2.7.: Comparison of a measured mode intensity profile (left) with a calculated one (right) in a 18-μm wide Ti-waveguide.

2.3.2. Nonlinear Conversion Efficiency

A typical wavelength scan with a 1064-nm pump is shown in figure 2.8 (measured values represented by dots). The particular waveguide has an overall length of 92 mm, of which 4 mm are lost due to the tapered input waveguide. The phase-match tuning bandwidth is about 1 nm in such a sample. In theory, a nearly mirror-symmetric curve resembling a $sinc^2$-type function is expected (broken line); however, waveguide inhomogeneities due to fabrication tolerances (waveguide width, crystal homogeneity, etc.) and temperature gradients in the waveguide oven can lead to a chirp in the coupling term, causing some asymmetries and reduced efficiency. Such a chirp has been modeled by adding a parabolic variation of the phase mismatch term $\Delta\beta$ (see equation 1.17) to the calculation, in order to re-create the slightly asymmetric measured curve. This is shown in figure 2.8 as a dotted line. Also, the effective interaction length was adjusted slightly for a better agreement between the curves. On the right side of the measured curve, a flat hump is discernible, which indicates additional linear and higher order terms in the chirp.

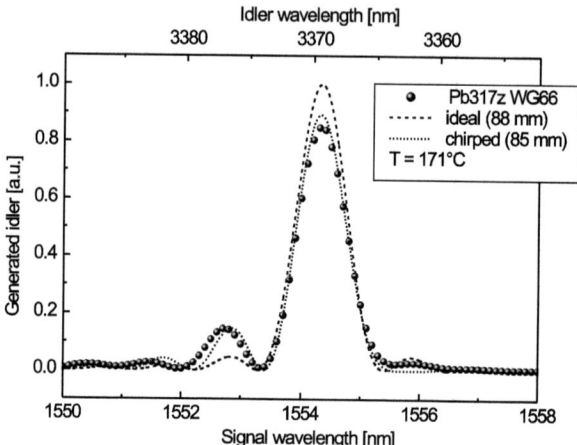

Figure 2.8.: DFG tuning characteristics. The generated idler is given as function of signal wavelength. Pump wavelength was fixed to 1064 nm. The difference-frequency wavelength of the generated radiation is given in the top axis. Dots represent the measurement; lines represent numerical calculations.

Assuming homogeneous crystal quality in a single sample, the asymmetries are likely caused by

- variations in Ti strip width,
- variations in Ti strip thickness,
- variations in the poling periodicity,

Parameter	Calculated variation	Measured variation
Ti strip width	-0.6 µm	-0.3 µm
Ti strip thickness	-1.72 nm	-1.1 nm
Poling periodicity	0.016 µm	n.a.
Indiffusion temperature	0.85°C	-0.5°C
Operating temperature	-7.4°C	n.a.

Table 2.1.: Calculated and measured variations of waveguide parameters along the sample length. The calculated values correspond to the phase-mismatch chirp that was determined to model the measured DFG tuning characteristics. The measured variation in Ti-strip width and -thickness, and indiffusion temperature of this particular sample compare well to the calculated values. In fact, due to the accumulation of effects in a real sample, it is reasonable that the measured values are somewhat smaller.

- temperature gradients in the indiffusion oven, and
- temperature gradients of the sample during operation.

The heuristically determined, parabolic chirp can now be traced to each of these causes. This was done by individually calculating the partial differentials of the phase-mismatch term $\Delta\beta$, as function of each. The result is given in table 2.1. Evidently, the measured Ti-strip parameters from the particular DFG sample compare very well to the calculated ones. The measured variations are slightly smaller even, due to the accumulation of effects in a real sample. The calculated temperature variation during indiffusion seems plausible; however, the temperature profile in the oven had a different sign in the variation. The necessary variation in operating temperature during operation is quite high, so the effect of temperature variations in the sample oven on the shape of the tuning characteristics should be accordingly small. Lastly, variations in poling periodicity are not known; however, it is assumed that the periodicity itself is quite homogeneous due to the fabrication process. See appendix C for some further discussion.

The low-power, normalized external conversion efficiency is defined as $\eta = P_{idler}/P_{pump}P_{signal}$. Pump and signal powers were measured before the WDM coupler (see figure 2.4) and determined as 32 milliwatts and 0.3 milliwatts, respectively, before being combined and coupled to the waveguide. At this power level, an MIR power of 1.2 µW was measured with an MCT detector. In this case, at a wavelength of 1553.4 nm, the external conversion efficiency was determined to be 13.5 %/W. A Germanium filter was used to block pump and signal, with a transmission of about 66 % at 3.4-µm wavelength. Also, fiber coupling losses need to be considered, which lie between 1.6 and 3 dB for each wavelength [46] (the exact coupling efficiency cannot be easily measured, mainly due to multimodal behavior of the waveguide at NIR wavelengths). The WDM coupler itself exhibits an insertion loss of 0.6 to 1 dB. From this we can conclude that the internal conversion efficiency lies between 58.5 %/W and 93.4 %/W, depending on

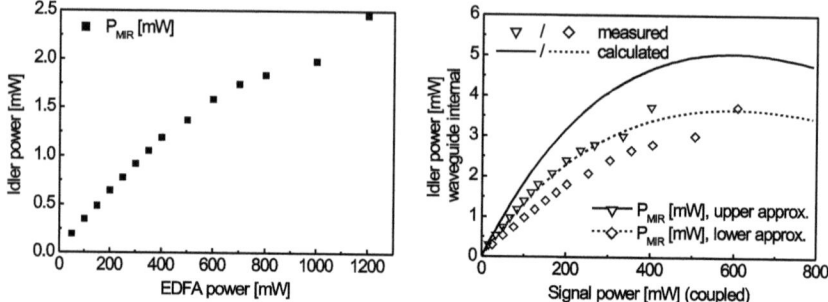

Figure 2.9.: DFG idler power dependence on signal power, with approximately 22 milliwatts of pump coupled to the waveguide. (a) Measured power levels. (b) Deduced internal power levels. Upper and lower limits were determined by varying fiber coupling coefficients between -1.6 and -3 dB, and -1.8 and -3 dB, for the pump and signal waves, respectively.

the assumed fiber-to-waveguide coupling losses.

The signal power dependence is shown in comparison to theory in figure 2.9. Pump power was kept constant (output power of laser diode of 32 milliwatts), with approximately 20 milliwatts coupled to the waveguide. Again, the internal power was estimated from the measurement. An upper and lower approximation of waveguide-internal powers was performed by varying the in-coupling coefficients between 1.6 to 3 dB for pump and signal radiation, and by considering the 0.6 to 1 dB insertion loss from the WDM as well as 33 % transmission loss of the pump-absorbing filter. Reflection losses from the sample were assumed to be low due to the AR-coatings. Power was measured using the MCT detector and 42 dB of attenuation using grey filters, in order to reduce the power levels to the linear regime of the detector. The grey filters were previously characterized at 3.4-μm wavelength.

The experimentally ascertained curves lie slightly below the calculated ones. A lower experimental conversion efficiency is even expected with respect to theory, due to the imperfections of the waveguide which manifests in the asymmetry seen in figure 2.8.

Further studies were performed using the high-power laser with 1 Watt of output power at 1064-nm wavelength. Approximately 400 milliwatts could be coupled to a fiber, however additional losses due to the use of a high-power fiber optical isolator and polarization control reduced the utilizable power to about 250 milliwatts. The setup used for power and polarization control as well as fiber coupling is shown in figure 2.10. An f = 5 mm ball-lens gave the best coupling efficiency without further beam shaping. Additional fiber-optic and coupling losses reduced the available power in the waveguide to roughly 80 milliwatts.

Using this pump source, a two dimensional power plot was measured by varying both, pump and signal power. The result is compared to the modeled pump and signal power

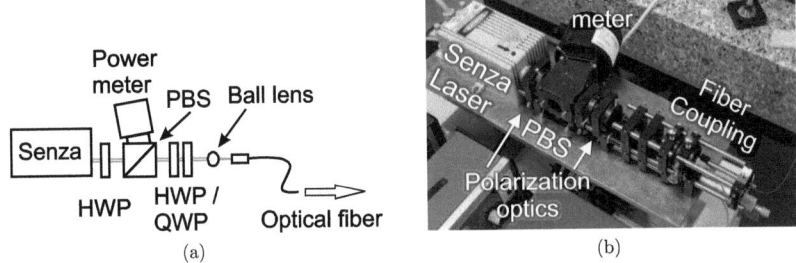

Figure 2.10.: Klastech Senza 1-Watt pump laser [44]. (a) Power and polarization control and fiber coupling scheme. A polarizing beam splitter (PBS) is used to control the output power by rotating the first half-wave plate (HWP). The second HWP, together with a quarter wave plate (QWP) are used to adjust polarization. (b) Corresponding photograph.

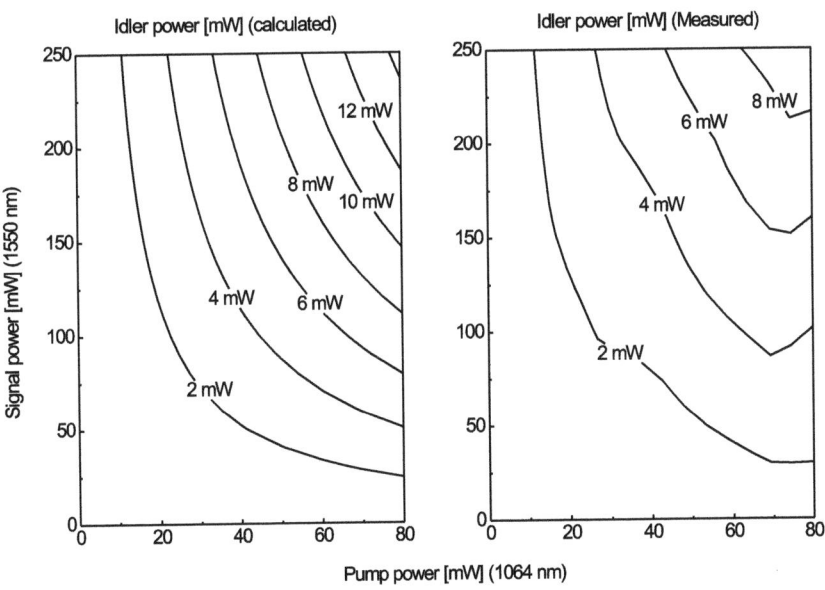

Figure 2.11.: Two-dimensional plot of generated idler power as function of coupled pump and signal powers. Left: calculated profile, right: measured profile with deduced waveguide-internal power levels.

dependence of the idler in figure 2.11.

2.3.3. Wavelength Characteristics

Sample temperature may be changed in order to tune the phase-matching wavelength in a waveguide of fixed periodicity. The change in temperature leads to a change of the phase-matching term (equation (1.17)), due to material dispersion according to a Sellmeier-equation [47] as well as waveguide dispersion, which was approximated by

$$n_{eff}(\lambda, T) = n_{LN}(\lambda, T) + \delta n(\lambda),$$

where the material dispersion for LiNbO$_3$ is considered by $n_{LN}(\lambda, T)$ and the waveguide mode dispersion by $\delta n(\lambda)$. Waveguide mode dispersion was evaluated using FOCUS [16] for the NIR and MIR ranges. The result, compared to the peak wavelengths of measured phase-matching curves, is shown in figure 2.12. It is demonstrated that using

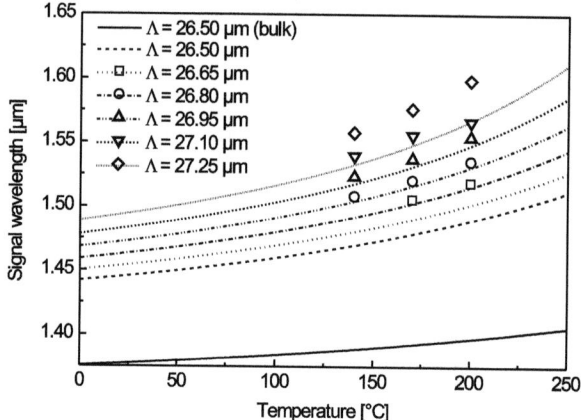

Figure 2.12.: Temperature dependence of DFG interaction due to waveguide dispersion, using a 1064-nm pump, in the 1550-nm signal wavelength range. The lines represent the calculated phase-matching evolution for different poling periodicities, while the symbols correspond to actual measurements.

a 1064-nm pump, the complete tuning range between 1500 and 1600 nm of a telecom-band ECL is accessible by choosing the appropriate poling periodicity and temperature. This corresponds to the generation of DFG-wavelengths in the range between 3176-nm and 3660-nm wavelength, using a 1064-nm pump.

The calculated phase-matching curves deviate by about 20 nm from the measured values at the given temperatures. Possible contributing factors to the deviations may be

- absolute temperature of the sample due to imperfect temperature controllers

- fabrication tolerances (e.g. waveguide width variations, grating inhomogeneities)
- absolute wavelength accuracy of the lasers used in the experiments
- imperfections of the waveguide dispersion model (the Ti-indiffusion profile, temperature dependence of waveguide mode dispersion, as well as the Sellmeier equation used to calculate material dispersion, all play a role).

It is, however, obvious that the general trend of the calculated curves is reflected in the measurements. The measurement in the waveguide with $\Lambda = 26.95$ μm exhibits an anomaly at 200°C. In retrospective, it may well be that a higher order mode combination was excited unintentionally. Calculations and previous measurements have shown that higher order mode combinations with significant mode overlap are always to be found at shorter wavelengths, compared to the fundamental mode combination [12].

2.3.4. Output-Power Stability

The temporal stability of the device was also examined. Stable MIR output power could be shown with a maximum of 3 % variation about the mean value during the first 1000 seconds in a measurement shown in figure 2.13. After 1000 seconds or so, however, output power reduces steadily. In the inset, an additional measurement is shown, which was conducted at another point in time. Here, a maximum variation of 4 % about the mean was observed over a period of 2800 seconds. This is a clear indication that the instability arises mainly due to mechanical vibrations and drift in the fiber coupling arrangement.

Additional contributions may arise from temperature variations in the sample as well as photorefractive effects, leading to variations in the phase-matching characteristics and conversion efficiency due to refractive index fluctuations in the waveguide. Clearly, amplitude noise of pump and signal sources directly contribute to the noise of the generated radiation.

The sensitive coupling condition, where radiation at two different wavelengths are coupled to an asymmetric waveguide with depth-shifted modes, leads to stringent requirements on mechanical tolerances. The stability measurement shows how critical a robust and stable fiber coupling mechanism is. In order to improve stability, one could think about:

- Fiber butt-coupling the ferrule to the end-face (possibly using high-temperature UV-cure and/or damage resistant waveguides at RT). This has not been done due to lack of adequate UV-cure. Possible damage resistant waveguides could be $LiNbO_3$ X-cut ridge waveguides, which were recently investigated by Gui [48].

- Using highly symmetric waveguides with symmetric and centered modes, with a higher translational tolerance of the mode overlap integrals and thus coupling efficiencies (see above).

- An active piezo fiber positioning system, using an output-power controlled feedback loop (a highly sophisticated setup would be necessary, however).

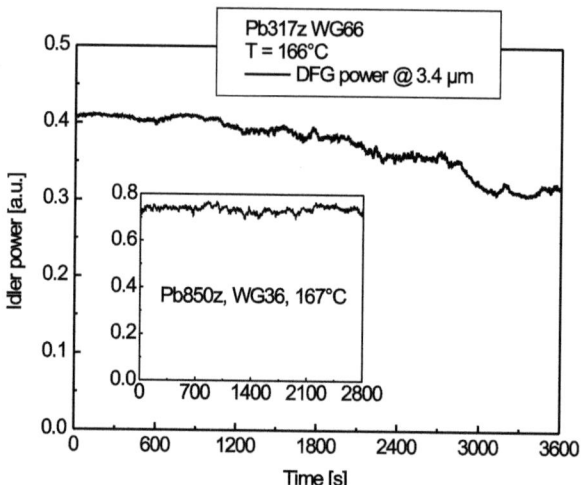

Figure 2.13.: DFG stability measurement. In the main graph, stable output was achieved for 1000 seconds, after which a degradation is observed due to thermo-mechanical instabilities. In the inset, 2800 seconds of stable operation was recorded at a different time.

2.4. Summary

Output powers above 10 milliwatts of 3.4-μm radiation and external conversion efficiencies above 10 %/W (normalized to the fiber-coupled pump power) were demonstrated by DFG in Ti:PPLN waveguides. The internal efficiencies are significantly higher, because waveguide-coupling losses reduce the external efficiency. The internal conversion efficiency, which was above 58 %/Watt, is comparable to what Asobe et al. demonstrated in Zn-doped ridge waveguides [29]. They reported 65 milliwatts of output with 444 milliwatts of pump, and 558 milliwatts of idler coupled to the waveguide, corresponding to a conversion efficiency of 26.2 %/W; however, reflection and filtering losses of the idler beam are not included, so the internal efficiency should be larger than 40 %/W in their case. Also, the waveguide was just 38-mm long with an estimated loss of 0.13 dB/cm in the MIR. This is a clear indication that indeed, the mode overlap of NIR and MIR modes is improved in ridge waveguides, compared to Ti-indiffused ones. They report a low-power conversion efficiency of 35 %/W and quantum efficiency of 46 %.

Instantaneous tunability of around 5 nm in the MIR range could be achieved, limited by the phase-matching bandwidth. The bandwidth may be increased by either shortening the interaction length of a sample (using shorter waveguides), or by use of an apodized grating, each at the cost of reduced conversion efficiency.

At 3.4-μm wavelength, the only conventional and commercially available cw-lasers are, to my best knowledge, Lead-Selenide (PbSe) laser diodes, so DFG is an attractive alternative. In fact DFG systems, based on bulk crystals [32] or waveguides [33] have recently been commercialized for the 3- to 4-μm wavelength range.

3. Hybrid Up-Conversion Detector

The potential of up-conversion of radiation and subsequent detection was recognized early on [4]. Bloembergen suggested to realize an infrared counter with an optical pump, to up-convert radiation using a 4-level solid state system.

The principle of the hybrid up-conversion detector (UCD) is the same, however nonlinear optical frequency conversion is used to convert radiation to shorter wavelengths, as depicted in figure 3.1. At the shorter wavelength, detectors with favorable properties like high sensitivity and fast response times are available (e.g. Silicon or InGaAs detectors).

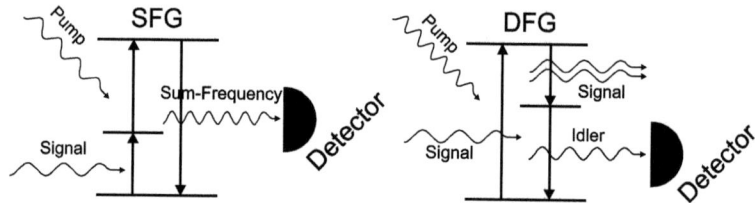

Figure 3.1.: Principle of SFG and DFG up-conversion detectors. The incident, long-wave radiation (signal) is up-converted by frequency mixing with a short wavelength pump. The generated photons are then collected by a suitable photo-detector.

In this chapter, a competitive hybrid up-conversion detector, using Ti:PPLN waveguides, and NIR detectors, is presented; building on the previous two chapters. Up-conversion using both SFG and DFG are demonstrated and experimental results, advantages and disadvantages are discussed in comparison to a conventional, HgCdTe (or MCT), MIR-detector.

In case of the UCD, some modifications of the wavelength conversion setup are necessary with respect to the DFG-MIR source, although the operation principles are largely identical. It should be clear that also the waveguide samples are identical in case the same combination of wavelengths is used, with the exception of an additional pump reflection mirror - which is, by the way, not detrimental to MIR-generation by DFG using the same sample. This means that, in principle, functionality may be exchanged and even transceiver modules are a possibility.

In practice, sample parameters may be optimized for specific wavelength combinations. Within the scope of this work, however, the wavelengths are fixed to the 1.1-μm, 1.55-μm, and 3.4-μm ranges. In this way, phase-matching can be achieved with the same samples for all three mixing processes used:

- In the case of the DFG-MIR source:
  $\omega_p - \omega_s = \omega_i$, $\Delta\beta = \beta_p - \beta_s - \beta_i - \frac{2\pi}{\Lambda}$, $\lambda_p = 1064$ nm, $\lambda_s = 1550$ nm, $\lambda_i = 3400$ nm

- In the case of the SFG-UCD:
  $\omega_p + \omega_s = \omega_{sf}$, $\Delta\beta = \beta_{sf} - \beta_s - \beta_p - \frac{2\pi}{\Lambda}$, $\lambda_p = 1550$ nm, $\lambda_s = 3400$ nm, $\lambda_{sf} = 1064$ nm

- In the case of the DFG-UCD:
  $\omega_p - \omega_s = \omega_i$, $\Delta\beta = \beta_p - \beta_s - \beta_i - \frac{2\pi}{\Lambda}$, $\lambda_p = 1064$ nm, $\lambda_s = 3400$ nm, $\lambda_i = 1550$ nm

3.1. Theoretical Background

In the following, detector figures of merit are introduced in order to characterize measured UCD performance, and to compare with conventional MIR detectors.

3.1.1. Basic Detector Figures of Merit

In case of the UCD, the overall (external) conversion efficiency determines the ultimate sensitivity of the detector system. The overall conversion efficiency defined by

$$\eta_{tot} = \tau_{opt}\eta_{nl} \tag{3.1}$$

relates the incident MIR power to the up-converted power which reaches the detector. Here, τ_{opt} describes the transmission through all optical components, including e.g. coupling efficiencies and fiber optic insertion losses through the spectral filtering apparatus. The nonlinear conversion efficiency within the waveguide ('internal' conversion efficiency) is given by η_{nl}. We can easily measure η_{tot} directly, while the internal conversion efficiency and optical transmission must be deduced from experimental data. Note that η_{tot} is a function of pump power, as η_{nl} is a function of pump power (at large incident MIR-powers, comparable to pump power, pump depletion may in principle also occur - I assume this is not the case). τ_{opt} necessitates a spectrally narrow input beam suitable for waveguide coupling.

We now can define for the hybrid detector

$$NEP_{HD} = \frac{1}{\eta_{tot}} \cdot NEP = \frac{1}{\eta_{tot}} \cdot \frac{S_n \Delta f^{1/2}}{R} \tag{3.2}$$

where

$$NEP = \frac{S_n \Delta f^{1/2}}{R} \tag{3.3}$$

is the noise equivalent power of the detector at the up-conversion wavelength (with pump enabled), and NEP_{HD} is the noise equivalent power in terms of incident MIR radiation. NEP is discussed in terms of the the noise spectral density S_n. R is the detector responsivity at the wavelength of interest. For characterization purposes, a lock-in amplifier with bandwidth Δf was used to determine S_n experimentally in terms of voltage, with $[R]$ = V/W.

The figure of merit which is usually used to describe the performance of a detector is the specific detectivity [49]

$$D^* = \frac{(A\Delta f)^{1/2}}{NEP}, \quad (3.4)$$

which is normalized to the detector area A and the measurement bandwidth, Δf. With equation (3.2), D^* becomes

$$D^* = \frac{R\sqrt{A}}{S_n}. \quad (3.5)$$

In order to determine D^*, S_n was measured using a Signal Recovery DSP7265 lock-in amplifier at a given measurement bandwidth Δf. D^* can then be calculated by plugging detector specifications into equation (3.5). In case of the hybrid detector, the measurements are conducted with the pump enabled, as residual pump radiation is a major noise contributor in a nonlinear-optical UCD. Thus, S_n is composed of intrinsic detector noise, noise from electric circuitry, and noise due to residual radiation.

3.1.2. Up-conversion Detector Performance Estimation

In order to estimate the ultimate performance of a UCD in comparison to a conventional detector, the ultimate specific detectivities D^* of MIR and NIR detectors are considered and discussed (using some standard detector theory), as well as the theoretical conversion efficiency η_{nl} of the nonlinear process.

Background-limited Detectivity

The following discussion is based mainly on [50]. At room temperature, MIR photo-detectors are limited by thermal recombination of charge carriers, which is why such detectors are usually cooled during operation. In case of a cooled MIR photo-detector, performance is limited by thermal background radiation due to the narrow band-gap. In case of a wider band-gap material, thermal radiation has less impact and performance is limited by other noise sources.

The noise current generated by a photo-detector is given by

$$\bar{i_n^2} = ue^2 G_b^2 A \Delta f \left[\int_0^\infty \eta(\lambda) E_{q,\lambda}(\lambda, T) d\lambda + \frac{k_B T_d}{e^2 G_b^2 r_0 A} + g_{th} d \right]. \quad (3.6)$$

The factor u equals 2 for photo-diodes and accounts for equal contributions of electron-hole generation and recombination to photo-current noise; in a photo-conductor, it is usually assumed as 4. The elementary charge is given by e. G_b is the background-dominated photogain which is set to unity in further considerations; $\eta(\lambda)$ the quantum efficiency, $E_{q,\lambda}$ the photon spectral background irradiance. k_B is the Boltzmann constant, T_d detector temperature, r_0 resistance and A detector area. The thermal generation rate is given by g_{th}.

The latter two terms describe Johnson noise and thermal generation, and will be neglected in further considerations. Johnson noise can be kept small in practice, by selecting a large value of the product $r_0 A$; and thermal generation may be minimized by keeping the device depth as small as possible without reducing quantum efficiency, according to [50]. The first term however describes noise due to background radiation and is assumed to dominate in a cooled MIR detector. It may be rewritten as

$$\int_0^\infty \eta(\lambda) E_{q,\lambda}(\lambda, T) d\lambda = \eta_0 \Phi_B(\lambda_c', T) = \eta_0 f_q(\lambda_c', T) \sigma_q T^3 \sin^2(\Theta), \tag{3.7}$$

with the fractional blackbody photon irradiance Φ_B incident on the photo-detector, an effective cut-off wavelength λ_c', and the asymptotic value η_0 of the quantum efficiency as function of wavelength. $\sigma_q = 1.5202 \times 10^{11}$ photons cm^{-2} s^{-1} is the photon equivalent of the Stefan-Boltzmann constant, and Θ is the solid angle determined by the field of view of the detector. $f_q(\lambda_c' T)$ describes the fraction of photons in the blackbody radiation field. In the case that $\lambda T \leq 5000$ μm K, it may be approximated in analytical form by [50]

$$f_q(\lambda T) = \frac{81}{\pi^4} \exp(-x) \left(1 + x + x^2\right), \tag{3.8}$$

with $x = hc/(\lambda k T)$ where h is the Planck constant. For the noise spectral density S_n follows

$$S_n = \sqrt{\overline{i_n^2}/\Delta f} = eG_b \sqrt{u\eta_0 \Phi_B(\lambda_c', T)}. \tag{3.9}$$

If we express the (current) responsivity by

$$R_i(\lambda) = \frac{e\lambda \eta(\lambda)}{hc}, \tag{3.10}$$

equation (3.5) may be re-written as

$$D^*_{BLIP}(\lambda, \Theta) = \frac{\eta(\lambda)\lambda}{hcu^{1/2}\left[\eta_0 \Phi_B(\lambda_c', T_b)\right]^{1/2}}. \tag{3.11}$$

BLIP stands for background-limited performance. An ideal detector has unit quantum efficiency. If we assume a 2π solid angle for the detector field of view, and room temperature background radiation, we get the typical curves displayed in figure 3.2. In case of a real detector, the wavelength dependent quantum efficiency must be considered, as well as additional noise contributors. Still, real infrared detectors approach the fundamental limit remarkably well [49].

Shot-noise - limited Detectivity

In case of photo-diodes in the NIR and visible wavelength ranges, the ultimate detectivity is dictated by the shot noise limit. The output current of such a device is

$$I = \frac{P}{h\nu}(1 - R)\eta e \tag{3.12}$$

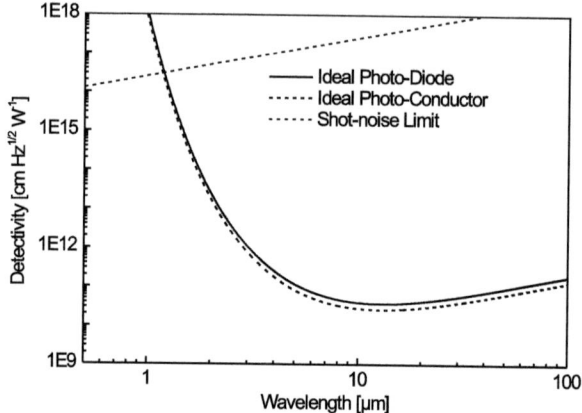

Figure 3.2.: Ideal detectivity of photo-conductive / photo-voltaic detectors. A 0.01 mm² detector at 1 Hz measuring bandwidth was assumed to determine D* for the shot-noise limited detectivity.

where P is the incident optical power, ν the optical frequency, e is the electron charge, R is the reflectivity of the detector surface, and η is the quantum efficiency. The mean square photon noise current is given by [51]

$$\bar{I}^2 = 2eI\Delta f \tag{3.13}$$

with a detection bandwidth Δf. In that case, the SNR can be expressed as

$$SNR = \eta(1-R)\frac{P}{2h\nu\Delta f}, \tag{3.14}$$

from which the NEP can easily be calculated as

$$NEP = \frac{2h\nu\Delta f}{\eta(1-R)}. \tag{3.15}$$

According to equation (3.4), the shot-noise limited D* becomes

$$D^*_{SN} = \sqrt{\frac{A}{\Delta f}\frac{(1-R)\eta}{2hc}}\lambda. \tag{3.16}$$

If we consider an ideal detector with unit quantum efficiency and neglect reflection losses, we get

$$D^*_{SN} = \sqrt{\frac{A}{\Delta f}\frac{\lambda}{2hc}}. \tag{3.17}$$

Though mathematically sound, this definition is not very useful per se, because the specific detectivity D* was introduced as a figure of merit for infrared detectors, normalized to the active detector area A and bandwidth Δf. In case of shot-noise limited performance however, the sensitivity is ultimately ascribed to the quantum nature of photons and electrons, which is inherently independent of detector properties.

Equation (3.17) is introduced here as a simple means to evaluate the fundamental performance limits of NIR detectors, compared to MIR detectors. Keep in mind these are for now hypothetical values, determined using simplified and fundamental theory.

Johnson Noise and Dark Current

NIR detectors are further limited by Johnson noise and dark currents (cp. equation 3.6). Johnson noise due to thermal agitation of charge carriers inside a detector with shunt resistance R_s is described by the variance [52]

$$\sigma_J^2 = \frac{4k_B T}{R}\Delta F. \qquad (3.18)$$

Taking a shunt resistance of 1 GΩ, which is a realistic value for a 100 μm diameter InGaAs photodetector [53], leads to a noise current in the range of 5 fA / Hz$^{1/2}$, which compares well to commercial InGaAs detectors. The NEP for a detector with roughly 1 A/W responsivity is thus reduced to about 5 fW/Hz$^{1/2}$.

Dark current due to random charge carrier generation in the depletion region leads to the term [52]

$$\sigma_D^2 = 2eI_0\Delta F. \qquad (3.19)$$

Assuming a dark current of 2 pA [54], this leads to noise currents in the range of 1 fA/Hz$^{1/2}$, which is below Johnson noise.

Using phase-sensitive detection, shot-noise limited performance may however be achieved [49].

Evaluation of Sensitivity Gain using Up-conversion

In table 3.1, detectivities taken from figure 3.2 are given at a number of interesting wavelengths, assuming either shot-noise limited performance or background-limited performance of photo-conductive and photo-voltaic photo-detectors. A relative gain, considering the limiting case at the particular wavelength (shot noise limit or background limit), is given relative to an ideal detector optimized for 3.4-μm wavelength.

A drastic improvement in background-limited detectivity is apparent if we go from the MIR to shorter wavelengths, assuming either thermal or shot-noise limited performance. At shorter wavelengths however, the shot-noise limit restricts the maximum detectivity to about 10^{16} cm$\sqrt{\text{Hz}}$ W^{-1}. In effect, a theoretical improvement between 10^3 to 10^5 can be expected when going to the shorter wavelengths. In reality, additional effects must

λ [nm]	$D^*_{SN}[\frac{\text{cmHz}^{1/2}}{W}]$	$D^*_{PD}[\frac{\text{cmHz}^{1/2}}{W}]$	$D^*_{PC}[\frac{\text{cmHz}^{1/2}}{W}]$	Relative gain [dB]
700	1.8 10^{16}	1.6 10^{22}	1.1 10^{22}	45
800	2.0 10^{16}	2.8 10^{20}	2.0 10^{20}	46
1064	2.7 10^{16}	2.9 10^{17}	2.0 10^{17}	47
1100	2.8 10^{16}	1.5 10^{17}	1.0 10^{17}	47
1550	3.9 10^{16}	5.1 10^{14}	3.6 10^{14}	30
3400	8.5 10^{16}	5.2 10^{11}	3.7 10^{11}	0

Table 3.1.: Comparison of background limited (300 K, $l > 1.1$ μm) / shot noise limited ($l < 1.1$ μm) specific detectivities at different wavelengths. SN: shot noise limited detector, PD: photo-voltaic detector (photo-diode), PC: photo-conductive detector (SN limited performance assumed if it is lower than the other values, cp. figure 3.2).

be considered, and detector geometry comes into play, as discussed before. So consider table 3.1 a rough theoretical estimation.

It is instructive, and in a way more sensible, to plot the noise equivalent powers according to shot noise and background limited detector theory, as the NEP of a shot noise limited detector is independent of detector geometry. See figure 3.3 for results. Shot noise limited performance could in principle be achieved below around 1-μm wavelength, depending on the detector geometry, and keeping the assumptions previously made in mind. However, Johnson noise restricts the performance of InGaAs detectors (at room-temperature) to the femtowatt-range per root bandwidth. At 1550 nm, an NEP of about 13.3 fW/$\sqrt{\text{Hz}}$ was measured experimentally with the Femto-InGaAs photodetector (using a lock-in amplifier, see section 3.2.2). With the MCT, an NEP of 14 pW/$\sqrt{\text{Hz}}$ was measured, again using lock-in technique (see appendix D for details on the MCT characterization). These are sensible values, because the InGaAs detector has a diameter of 1/10 of the MCT detector. Assuming a limitation by Johnson-noise in the InGaAs detector, and considering the different detector areas, this translates to about a factor of 100 difference in detectivity.

According to equation (3.2), the total conversion efficiency η_{tot} must be known in order to estimate the performance of a UCD detector. Consider the case of a lossless UCD with optical transmission τ_{opt} of unity. We need to differentiate between SFG and DFG:

- In the case of SFG, maximum conversion efficiency (absolute) is reached when 100 % of all incident MIR photons are up-converted by mixing with pump radiation. In terms of power, the ultimate conversion efficiency is given by the relation of incident and up-converted photon powers:

$$\eta_{tot,max} = \frac{h\nu_{SF}}{h\nu_{MIR}} = \frac{\nu_{SF}}{\nu_{MIR}} = \frac{\lambda_{MIR}}{\lambda_{SF}}.$$

Taking $\lambda_{MIR} = 3.4$ μm and $\lambda_{SF} = 1.064$ μm, this corresponds to a maximum conversion efficiency of 319 %.

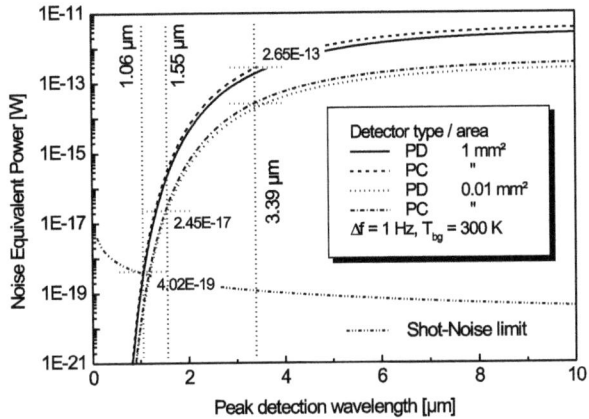

Figure 3.3.: Comparison of noise equivalent powers [NEP] of shot-noise / background limited detectors. Different detector geometries were assumed for background-limited detectors (see inset). Below 1-μm wavelength, the contribution from thermal background radiation drops below the shot-noise limit.

- In the case of DFG, parametric gain is accessible with sufficient pump power. For the time being, considering 100 % photon conversion efficiency, the power conversion efficiency is given by

$$\eta_{tot,max} = \frac{h\nu_{DF}}{h\nu_{MIR}} = \frac{\nu_{DF}}{\nu_{MIR}} = \frac{\lambda_{MIR}}{\lambda_{DF}}.$$

With $\lambda_{DF} = 1.55$ μm this equates to 219 %. At higher pump powers, conversion is enhanced due to parametric amplification and limited only by physical constraints. However, parametric fluorescence is also generated at the up-conversion wavelength, which adds a DC offset to the signal, as well as parametric noise [55].

Conversion efficiency was again calculated as function of pump power for both cases, considering waveguide losses as well as pump depletion, by numerical integration of the coupled mode equations (1.11) (figure 3.4. The initial MIR power was set to 150 μW, with an assumed coupling loss of -1.5 dB. The calculated efficiency peaks at 246.5 % at 600 milliwatts of pump in case of SFG - it is smaller than the unit conversion efficiency of 319 % due to the waveguide losses. In case of DFG, considerable gain is evident at coupled pump powers above 500 milliwatts and at 1 Watt of coupled pump, the conversion efficiency surpasses 1000 %.

In reality, conversion efficiency is likely to be reduced due to waveguide inhomogeneities (e.g. width and depth along the interaction length) and inhomogeneities of the periodic poling grating (e.g. varying duty cycle) due to fabrication tolerances, as well as

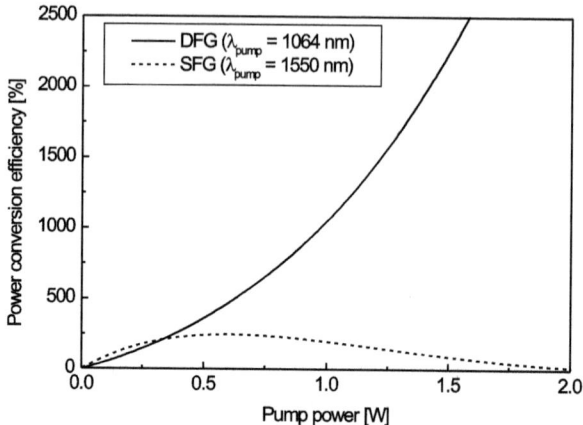

Figure 3.4.: Calculated UCD conversion efficiencies (SFG/DFG), including losses. Theoretically, a few hundred milliwatts pump power is sufficient for 100 % power conversion efficiency. In the DFG case, strong parametric gain is apparent when going to high pump powers.

photo-refractive damage effects (e.g. loss of phase-matching) at high pump powers. In addition, the external conversion efficiency is reduced due to coupling losses and other loss mechanisms.

Now, assuming a sensitivity improvement by two orders of magnitude (assuming a Johnson-noise limited InGaAs detector), and 100 % nonlinear quantum efficiency, a maximum improvement by a factor of about 250 can be expected using an SFG-UCD compared to an MIR detector.

3.2. Up-Conversion Detector based on Sum-Frequency Generation

In this chapter, the characterization of the up-conversion detector using Ti:PPLN waveguides, which is based on SFG, is described in detail. A photo of the setup which was used for characterization is shown in figure 3.5. The photo is applicable to the characterization of both SFG and DFG up-conversion detectors.

3.2.1. Experimental Setup

A schematic of a UCD exploiting SFG is shown in figure 3.6. Here, 3.4-μm radiation impinging on the device shall be up-converted to the NIR. Bear in mind that only linearly

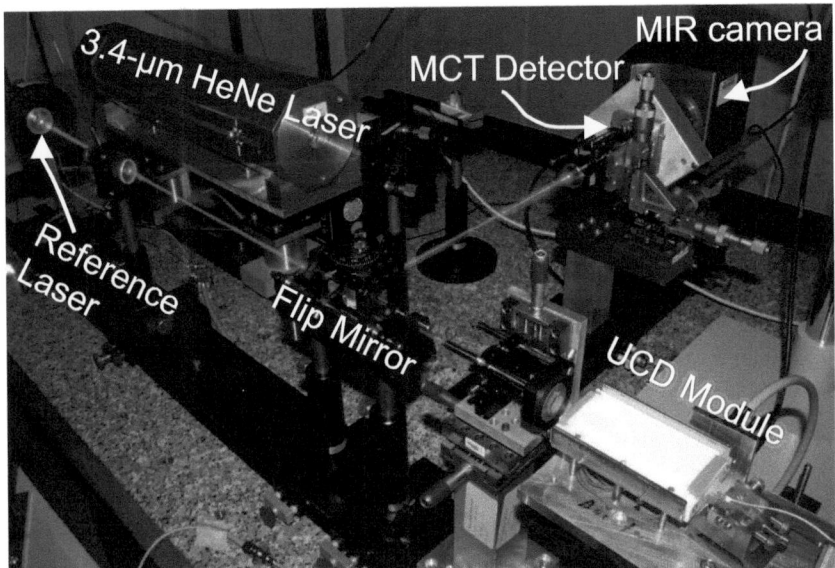

Figure 3.5.: Experimental setup for UCD characterization. The HeNe-laser used as MIR source, a reference laser diode, and the HgCdTe detector and MIR camera, which were used for further characterization and evaluation of the modules, are visible. The beam paths are illustrated by the red reference laser beam.

polarized light can be up-converted due to the polarization sensitive conversion.

In comparison to the DFG-MIR source, we need to change the following:

- In this case we are mixing radiation of 3.4-μm wavelength with 1550 nm radiation, using the ECL together with an EDFA as pump, to generate the SF at 1064 nm wavelength. The 1064 nm pump of the previous setup is replaced by an NIR photo detector.

- While the MIR beam is coupled into the waveguide from the free-space input, the pump is again coupled from a fiber. Hence, a pump reflection mirror is needed to reflect the pump at the free-space end face to enable co-propagating SFG.

- The SF-signal is routed through the 1064 nm port of the WDM to the detector.

- In order to achieve high signal-to-noise ratios (SNR), spectral filtering of the sum-frequency signal before it reaches the detector is crucial.

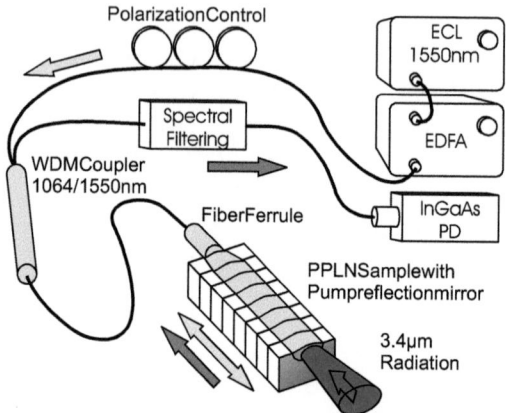

Figure 3.6.: SFG up-conversion detector scheme. Pump radiation from the tunable ECL laser is amplified by the EDFA, in order to up-convert the 3.4-µm radiation which is coupled to the waveguide. The generated SF is routed to the InGaAs photodetector (PD). The pump reflection mirror is indicated by a color gradient.

3.2.2. Results & Discussion of Detector Performance

In order to characterize the UCD for sensitivity (detector bandwidth shall not be taken into consideration at the moment), according to equation (3.2), the overall conversion efficiency η_{tot}, which includes the internal conversion efficiency of the waveguide η_{nl} as well as overall transmission τ_{opt} of the hybrid detector components, must be determined. η_{tot} is given by the ratio of up-converted radiation reaching the photo-detector of the UCD and MIR-power in the free-space beam, which was measured using the MCT detector. The characterization setup is shown in figure 3.8a.

The photo-detector itself is a variable gain InGaAs PIN photo-receiver (type Femto GmbH OE-200-IN2 [54]), with a maximum gain of $G = 10^{11}$ V/W at 1550 nm. Its best noise performance is achieved at the maximum gain setting; the upper cut-off frequency in this case is 1.2 kHz. The noise spectral density spectrum S_n of the detector was measured with the lock-in amplifier (figure 3.7). The responsivity of the detector must also be considered, because the same detector is used for different wavelengths, depending on the up-conversion process (see also section 3.3). In the SFG-UCD case, the generated radiation has a wavelength of 1064 nm and the responsivity is reduced to 73% of the calibration value at 1550 nm wavelength.

The lock-in noise measurement reveals a noise floor of about 0.6 mV/$\sqrt{\text{Hz}}$, corresponding to a spectral noise density of $S_n = 13.3$ fW/$\sqrt{\text{Hz}}$ at 1550 nm wavelength. At 1064 nm, the reduced responsivity of the detector must be taken into account, which gives $S_n(1064nm) = 18.2$ fW/$\sqrt{\text{Hz}}$.

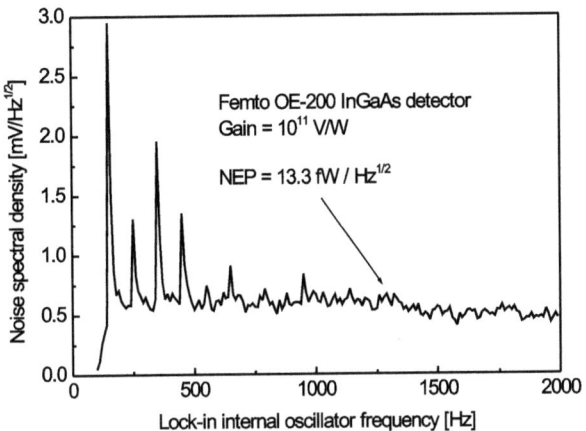

Figure 3.7.: Femto OEC-200-IN2 noise spectrum at 1550-nm wavelength as function of frequency. The spectrum was recorded by scanning the reference frequency of the Lock-In amplifier, using the internal oscillator, and measuring the dark voltage of the detector.

The UCD experimental setup scheme is shown in detail in figure 3.8b. A HeNe-laser is used as the MIR source at 3.39-μm wavelength. The light is coupled into the waveguide using a f = 15 mm ZnSe bestform lens. Two 1064 nm / 1550 nm WDM couplers are used to separate the up-converted signal from pump radiation with an isolation ratio of -20 dB each. The insertion loss of the WDM couplers is about 0.6 dB each, in addition to fiber connector losses. A 1064-nm optical circulator, used in conjunction with a fiber-Bragg-grating (FBG) centered at 1065 nm, is used to further isolate the sum-frequency signal with more than 57 dB isolation from the pump. Moreover it will eliminate spurious second-harmonic radiation of the pump at around 775 nm. The signal losses due to the filtering amount to -3.7 dB.

In order to enable co-propagating SFG in the waveguide, a dielectric mirror was deposited onto the end-face of the sample. It consists of a Monte-Carlo optimized 14-layer system of TiO_2 and SiO_2 layers. The two transmission dips at the pump wavelengths for both, SFG- and DFG-UCD, are shown in figure 3.9. The curve was measured using a $LiNbO_3$ witness sample.

The overall (i.e. external), *low-power* conversion efficiency (i.e. in the linear regime of conversion) with sample Pb317z was found to be $\eta_{norm} = 44.5 \frac{\%}{W}$ at 166°C sample temperature and 1553.7-nm pump wavelength, determined at an incident power level of 170 μW coming from the HeNe laser and just 1.15 milliwatts of pump before the WDM coupler (wavelength filtering *not* used in this case). When the additional insertion losses of the WDMs (ca. 0.6 dB plus fiber connector losses), coupling losses (about 1.8 dB at 1550 nm, about 1.6 dB at 1065 nm and about 1.3 dB at 3.4 μm), as well as pump

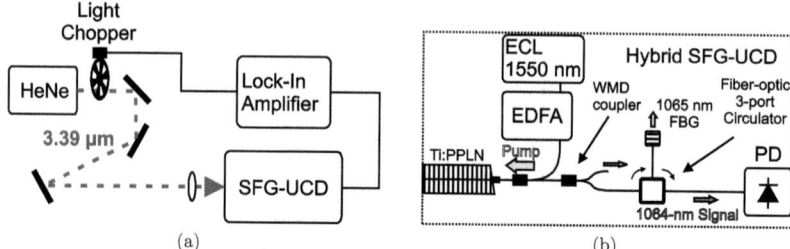

Figure 3.8.: (a) Characterization setup used to investigate the SFG-UCD. The three mirrors were used to steer the MIR beam for alignment purposes. (b) Detailed view of the SFG-UCD, including the fiber-based wavelength filtering scheme.

propagation and mirror losses (ca. 1 dB) are considered, this yields a normalized, internal conversion efficiency of $\eta_{nl,norm} = 262 \frac{\%}{W}$. The uncertainty is quite large, at about 3.5 dB, mainly due to coupling uncertainties.

At higher pump powers, the *normalized* conversion efficiency decreases. The measured power dependence of converted power is shown in figure 3.10. In the left figure, the external power is given, again adjusted for pump filtering losses (not considered here). In the right figure, the waveguide-internal power levels were deduced. The coupling coefficients were varied between -1.8 and -3 dB for pump, and -1.6 and -3 dB for sum-frequency (signal) radiation, and between -1.3 and -5 dB for the MIR beam, in order to approximate lower and upper bounds for the internal conversion efficiency. This considerable variation of the coupling coefficients was assumed because multi-mode behavior of the waveguide, the taper structure, and the mirror, as well as instabilities in the modal structure of the HeNe laser, all contribute to complicating the determination of the 'true' coupling coefficients for an individual waveguide sample.

A marked reduction in conversion efficiency is evident at higher power levels, which is in principle in agreement with theory; however, instability effects (especially of pump polarization with EDFA power) could lead to a slight distortion of reality. Also, at pump powers considerably above 500 milliwatts, a pronounced deterioration of efficiency was observed over short periods of time (within seconds) - after reducing pump power back to about 500 milliwatts, the efficiency was recovered within a comparable timescale. This is a clear indication of an 'optical damage' effect due to photo-refractivity [28].

With the aforementioned pump filtering scheme, the additional noise contribution due to pump radiation and the second harmonic thereof was found to be negligible. Fiber-optic insertion losses imposed by the filtering components were determined as 3.7 dB due to the circulator plus FBG, and 1 dB due to the additional WDM. Plugging the noise spectral density $S_n(1064 \text{ nm}) = 18.2 \text{ fW}/\sqrt{\text{Hz}}$ and taking the total conversion efficiency $\eta_{tot} = 6.15\ \%$ at 1 Watt of pump into equation 3.2 yields $NEP_{HD} = 0.3$ pW at one Hertz measurement bandwidth. In comparison to the MCT detector (see appendix D),

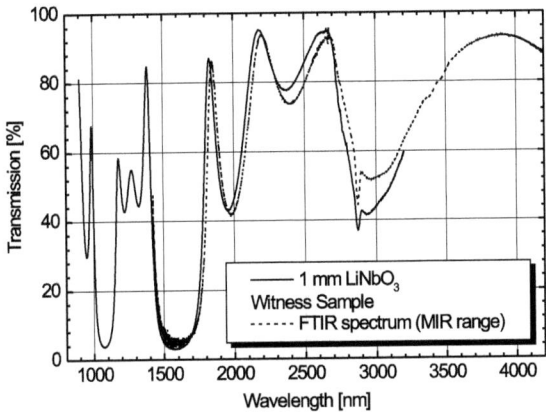

Figure 3.9.: UCD pump-reflection end-face mirror characteristics. The two transmission dips at 1064 nm (R = 96.1 %) and 1550 nm (R = 96.8 %) are visible. Transmission at 3.4 μm may still be improved with T = 76.2 %, according to the FTIR spectrum. Spectra were measured using a witness sample. Note: the FTIR spectrum was recorded 4 months later than the grating spectrometer measurement, with an apparent slight change in transmission properties.

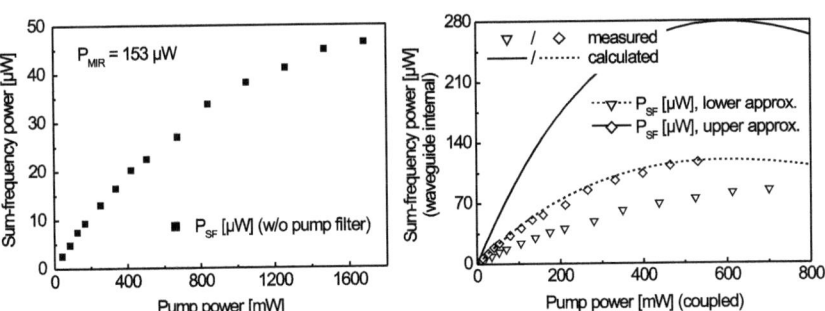

Figure 3.10.: Left: SFG-UCD power characteristics. Right: Corresponding waveguide-internal conversion compared to theory, with lines representing calculations. Varying waveguide coupling coefficients (free-space and fiber coupling) were used in order to approximate upper and lower boundaries for both experimental and calculated power values. A fairly high variation (-1.3 to -5 dB) of the free-space coupling coefficient of MIR-radiation was assumed, due to modal instabilities of the HeNe-laser.

this is an improvement by a factor of 48; however, the differing detector areas must also be taken into account. The MCT has an active area of 1 mm² while the fiber-coupled InGaAs detector has an area of just $7.85 \cdot 10^{-3}$ mm². The 'effective' detectivity of the hybrid detector may now be defined as

$$D^*_{eff} = \eta_{tot} D^* \qquad (3.20)$$

and equates to $3.0 \cdot 10^{10} \frac{\text{cm}\sqrt{\text{Hz}}}{\text{W}}$. In the case of the MCT, we had determined $D^* = 7.14 \cdot 10^9 \frac{\text{cm}\sqrt{\text{Hz}}}{\text{W}}$. The effective improvement is thus reduced from 48 by a factor of 11.3, so the (arguably) 'real' improvement of the hybrid detector over the MCT detector is about 4.2 times higher sensitivity, using the setup shown (assuming background-limited detector theory). Frankly, due to the acceptance angle of the waveguide coupling, and due to the spectrally selective phase-matching process, there is some spectral and spacial filtering going on which could, frankly speaking, be realized otherwise with a conventional detector. If the waveguide mode size is taken as the reference detector area, the SFG-UCD even performs by about a factor of 2 worse than the MCT in terms of sensitivity (albeit using the InGaAs detector with a much larger area).

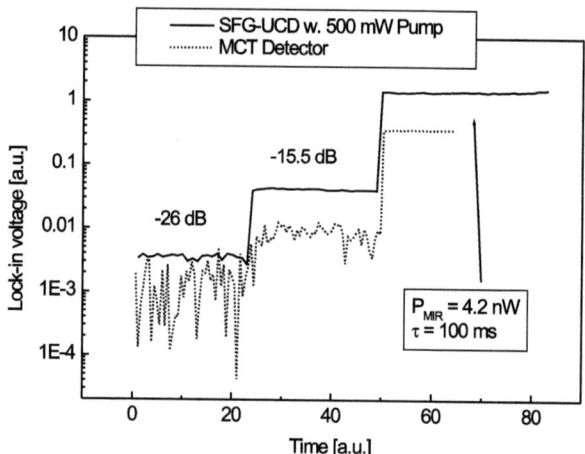

Figure 3.11.: Qualitative comparison of the MCT detector and the SFG-UCD. A strongly attenuated MIR beam from the HeNe laser was measured with each detector. The attenuation was chosen so that the signal dropped below the noise level of the MCT, and was subsequently reduced. In case of the MCT, the absolute value was taken for plotting purposes, as the noise oscillates around zero (signal amplitude is arbitrary).

Still, in a direct comparison, the UCD performs remarkably well. In figure 3.11, a time-resolved measurement of an MIR signal is shown at three different attenuation strengths. Starting at the highest attenuation in this case, a clear signal can be detected by the SFG-UCD, while the MCT sees only noise (in fact, the absolute value was taken in case of the MCT to allow the logarithmic scaling, because the voltage oscillated around zero). With 11.5-dB less attenuation, the MCT sees the signal, which is still significantly more noisy than the UCD one. With an additional 15.5-dB less attenuation, both signals are essentially noise-free. A very slight drift is discernible in case of the UCD, due to lower mechanical tolerance (i.e. waveguide coupling) in that case.

There is a potential of about 4.7 dB improvement of detector performance by replacing the lossy fiber-optic components with an ideally lossless wavelength filtering scheme (cp. figure 3.12). If the waveguide-fiber coupling losses could be eliminated - either by using

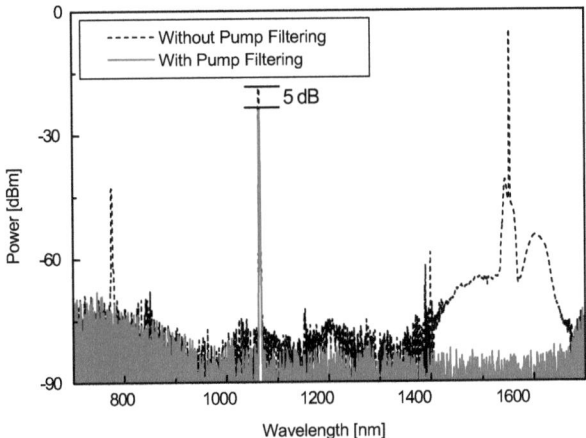

Figure 3.12.: Optical spectrum measured at the 1064-nm output port of the WDM coupler, measured with and without spectral filtering components, of the SFG-UCD. Without filtering, residual pump radiation with a broad pedestal is evident around 1500 nm; also, a second harmonic peak at 775 nm is evident. The ca. 5 dB excess loss due to the filtering components is evident. Incident MIR power was 150 μW.

an optimized bulk optic coupling scheme, or by employing highly symmetric waveguides - an additional 1.6 dB could be gained in out-coupling of the sum-frequency, and 1.8 dB of available pump power could be gained. Also, the coupling coefficient of a TEM_{00} MIR beam, as well as the nonlinear conversion efficiency could be improved by using more symmetric waveguides. Assuming that the direct comparison of the UCD and the MCT is valid, this would mean that a sensitivity improvement by a factor of more than 200 would be reached, with less external pump power. In the 'worst case scenario' (taking the mode size into account as the active detector area), an improvement by a factor of 2 would be achieved. Also, pulsed operation with high peak powers and thus high conversion efficiency is a possible method to alleviate the problems with high pump powers and corresponding optical damage effects. Optical damage-resistant waveguides, using e.g. Mg- or Zn-doped [29, 56], damage-resistant LiNbO$_3$ would be another option for high-power operation. Asobe et al. have reported up to 46% quantum efficiency in case of MIR-generation in such a ridge waveguide [29]; this would mean that the theoretically predicted improvement would even be in reach with such waveguides. Finally, the use of an alternative NIR-detector, e.g. a thermo-electrically cooled detector based on InGaAs, could be used in the UCD to improve performance further.

Krier and Mao have demonstrated a detector based on InAsSbP on InGaAs [57], sensitive in the 1.6- to 3.4-μm wavelength region. They published a sensitivity of $1 \cdot 10^{10} \frac{\text{cm}\sqrt{\text{Hz}}}{\text{W}}$ at 3.2 μm wavelength at room-temperature. Still, the current SFG-UCD performs well in comparison. Temporão et al. [58] have investigated an SFG-UCD to up-convert nanosecond pulses at 4.3-μm wavelength, using a 980-nm pump, to 800 nm using a bulk PPLN-crystal. With a nonlinear conversion efficiency of just $\eta_{SFG} = 5.06 \cdot 10^{-3}\%$, and using a Si-APD, they saw a sensitivity improvement by a factor of 180 compared to a liquid-nitrogen cooled MIR detector (phase-sensitive detection not used). The MIR detector was optimized for high modulation bandwidths - as opposed to ultimate sensitivity - with a timing resolution of 20 ns and an NEP of 223 pW. Still, the Si-APD would outperform with 0.3 ns timing resolution, using PPLN up-conversion.

3.3. Up-conversion Detector based on Difference-Frequency Generation

In this chapter, the characterization of the up-conversion detector using Ti:PPLN waveguides and based on DFG is described in detail.

3.3.1. Experimental Setup

The experimental setup may be modified slightly to realize a DFG-based UCD. The scheme is shown in figure 3.13:

The changes compared to the SFG-UCD are as follows:

- The device is now pumped using a 1064-nm laser. The incident 3.4-μm-radiation mixes with the pump to generate a DF-signal at 1550 nm.

- The DF-signal is routed to the 1550-nm-port. After spectral filtering, it is detected using the same photo detector as before.

The characterization of the DFG-UCD is analogous to that of the SFG-UCD in section 3.2.2. The experimental setup is mostly identical to that of the SFG-UCD, however, the wavelength filtering scheme is adapted to the exchanged wavelengths of pump and up-converted radiation. In figure 3.14b, the scheme is displayed. Here, a fiber-optic circulator operating at 1550 nm, together with a (tunable) fiber-Bragg-grating with a

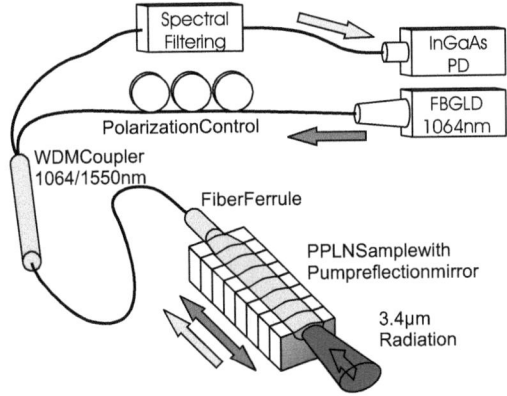

Figure 3.13.: DFG up-conversion detector scheme. Pump radiation in this case is fixed to 1064 nm. The DF at 1550 nm, generated from the incident 3.4-μm radiation, is routed to the InGaAs photodetector (PD). The pump reflection mirror is indicated by a color gradient.

center wavelength around 1550 nm are used in order to separate the up-converted radiation from residual pump radiation at 1064-nm-wavelength (and the second harmonic thereof at 532 nm).

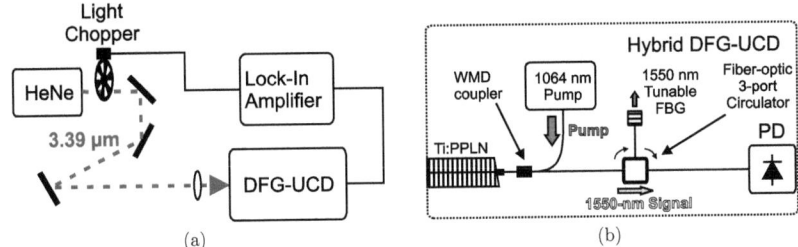

Figure 3.14.: (a) DFG-UCD characterization setup. (b) Detailed view of DFG-UCD, including wavelength filtering scheme.

3.3.2. Results & Discussion of Detector Performance

Tunable pump power of up to 230 milliwatts at 1064-nm wavelength was available from the Klastech laser [44]. The laser output power characteristics had to be thoroughly characterized. The pump power dependence of the UCD, again both external (not considering wavelength filtering) as well as the deduced internal power levels along with corresponding calculations, are shown in figure 3.15.

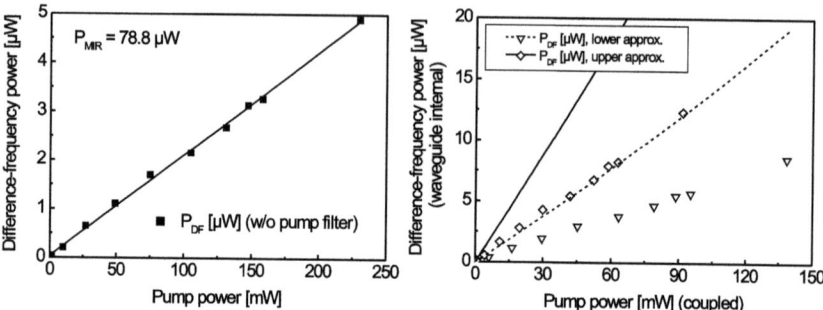

Figure 3.15.: Left: DFG-UCD power characteristics. Right: Corresponding waveguide-internal conversion compared to theory, with lines representing calculations. The same assumptions hold as for figure 3.10.

The onset of parametric amplification is not evident. Compared to the SFG-UCD, at least the conversion scales approximately linearly with pump power. Conversion efficiency was calculated as well and the results are depicted in figure 3.16. At pump powers above a few hundred milliwatts, the absolute conversion efficiency increases exponentially due to the contribution of the additional MIR-idler photons generated during the up-conversion process by DFG.

Experimentally, a conversion efficiency of 6.24% at 230 milliwatts of external pump, or 26 $\frac{\%}{W}$ normalized to pump power, was achieved (pump filtering not considered). The fiber optic pump filtering scheme adds an additional signal loss of 3 dB. At the highest pump power levels, parametric fluorescence [11] around 1550 nm was observed using an optical spectrum analyzer, with power spectral density within the range of hundreds of picowatts per nanometer. The measured fluorescence is strong enough to clearly resolve the phase-matching characteristics of the device (figure 3.17).

The fluorescence ultimately limits the performance of the detector, because it stems from the same phase-matched interaction which is responsible for up-conversion and detection. Consequently, it cannot be filtered spectrally - as opposed to pump radiation - and thus adds a permanent DC floor to any up-converted signal. The DC-component may be disregarded when using AC-coupled detection, however fluctuations in the fluorescent power adds additional noise even to the AC signal.

Still, a NEP of 0.6 pW could be achieved at 1 Hz measurement bandwidth, which is about a factor of 2 worse than that of the SFG detector. In direct comparison to the MCT detector, this is still an improvement by a factor of 23.3. The improved sensitivity was directly verified by attenuating the beam below the NEP of the MCT, at which power level it could still be clearly seen by the hybrid detector. Taking the different detector areas into account, in analogy to the considerations in the previous chapter,

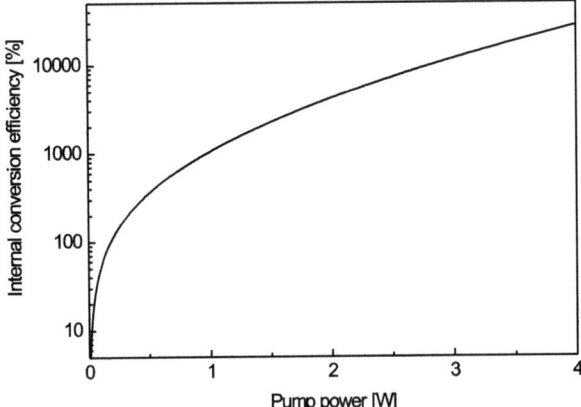

Figure 3.16.: High-power calculation of conversion efficiency in the DFG-UCD. At pump powers above a few hundred milliwatts, conversion efficiency proliferates due to parametric amplification.

the detectivity gain reduces to a factor of 2. Again, if the mode size is taken as the reference detector area, the UCD performs worse than the MCT detector by about a factor of 5, in terms of sensitivity.

3.4. Additional Considerations

If we compare both versions of the hybrid UCD detector, namely SFG-UCD and DFG-UCD, respectively; a number of factors should be considered. Keep in mind that the waveguide samples used are identical, only the setups differ. Therefore, this comparison concentrates on the aforementioned wavelength combinations. Factors which need to be considered are:

1. Target wavelength of the up-converted signal,

2. detector sensitivity / responsivity at the target wavelength,

3. pump wavelength and power and implications for the up-conversion process,

4. quality of spectral filtering.

The wavelength of radiation, which is to be up-converted, can be modified by tuning the pump wavelength - in principle, scanning of the pump wavelength can even be used for spectral characterization of radiation, which has been demonstrated in the 1310 nm range [59]. However, the phase-matching curve should be sufficiently narrow to achieve

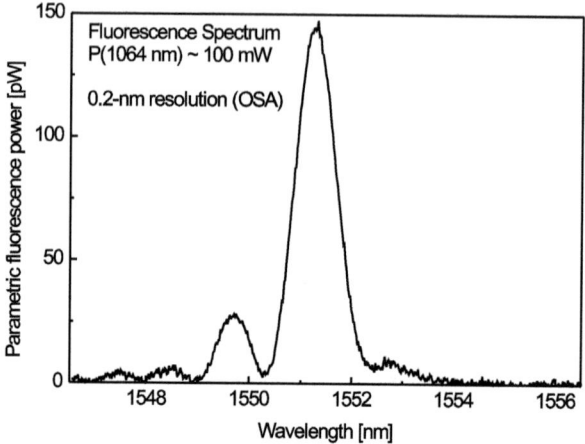

Figure 3.17.: Parametric fluorescence spectrum of the DFG-UCD at 1550 nm, measured at the output of the UCD with an Optical Spectrum Analyzer (OSA). Harmful parametric fluorescence is observed in the DFG-UCD only, because it generates the same phase-matched wavelength combination which are used in the DFG-process.

ample spectral resolution.

The target wavelength (SF/DF) may be optimized for a certain detector type (e.g. so-called single photon detectors, like APD's, could be used), so that optimum sensitivity is achieved. However, other aspects should also be considered. The mode overlap, which contributes to conversion efficiency, worsens if the wavelengths involved are further apart. Also, efficient and selective coupling to the fundamental modes is necessary at all wavelengths involved in order to facilitate an efficient interaction as well as efficient coupling to a single-mode fiber. Photorefraction needs to be considered especially when short pump wavelengths are used (e.g. pumping in the visible range) [28].

At the desired pump wavelength, adequate radiation sources should be available, and fiber coupling should be feasible for a waveguide based UCD. Output power should be sufficiently high to enable a competitive conversion efficiency. In the SFG case, an EDFA can be used to boost external pump power to well above 1 Watt. The power available from the 1-Watt, 1064-nm pump for the DFG experiments was limited to below 300 milliwatts due to fiber-optic coupling losses.

In order to measure a clean signal at the up-converted wavelength, residual radiation (e.g. pump radiation and the SH thereof) needs to be eliminated before reaching the detector. In the experiments described, an InGaAs photodetector[1] was used. The

---

[1] The Femto GmbH OE-200-IN2.

detector shows a relatively flat response characteristics between 1000 nm to 1600 nm, with peak responsivity at 1600 nm wavelength. At 1064 nm wavelength, responsivity is reduced to 73% of the maximum. Mainly the pump radiation, as well as the SH of the pump, deteriorate detector performance severely if they are not removed. The effort to spectrally separate the up-converted signal by a filtering scheme should be reasonable. Essentially two different fiber-optic filter concepts were used. A fiber-optic circulator together with a narrow-band fiber Bragg grating (FBG) is one option to isolate a specific wavelength, typically with 60 dB suppression of residual radiation. Also, WDM filters with around 20 dB - 25 dB suppression of either design wavelength can be used. AWGs (arrayed waveguide gratings) are another possible option.

At first glance, the DFG-UCD has the additional advantage of parametric amplification, which is inherent to the DFG-process. However, parametric fluorescence is also a phase-matched process and adds a permanent noise-floor. Parametric amplification as well as fluorescence are absent in the SFG-UCD. Also, the initial power-conversion efficiency is higher, due to the generation of higher power photons. The responsivity of the InGaAs detector is, however, slightly worse in the SFG-UCD due to the higher power photons.

3.5. Summary

In the preceding sections, up-conversion based detectors for MIR radiation at 3.4-μm wavelength have been presented using both SFG and DFG. For a given MIR wavelength, the phase-matched wavelength of the up-converted signal is dictated by the process (SFG or DFG) as well as the pump wavelength. To recapitulate, in the case of SFG, a pump with wavelengths in the 1550 nm range was used to generate the SF signal at around 1064 nm. In the case of DFG, a pump with 1064 nm wavelength was used to generate the DF signal at around 1550 nm.

With the SFG-UCD, a sensitivity improvement by a factor of 48 was seen experimentally, by directly comparing the UCD to a reference MCT detector optimized for 3.4 μm. If the detector area is taken into account (which was done by convention), the advantage reduces to a factor of 4.2. Similarly, the DFG-UCD achieved a sensitivity improvement by a factor of 23.2, which reduced to a factor of 2 by considering the smaller size of the InGaAs detector. If the waveguide mode size is taken as the reference detector area, the UCD's actually perform worse than the MCT. This was concluded based on theoretical considerations; the ultimate test would be to compare a MIR-detector with a detector area comparable to the mode size of the waveguide, to the UCD detectors. Such an MIR detector was, however, not available.

The main reason for the improved sensitivity in direct comparison is given by the fact that the incident radiation is up-converted with a high spectral selectivity, so background radiation poses less of a problem. However, additional advantages of the UCD-detector are:

- Easy interfacing to fiber optic components (e.g. detectors and modulators) and

- availability of detectors with superior temporal performance (i.e. higher modulation bandwidths are possible) than MIR-detectors.

The hybrid UCD may still be improved further. Main issues are the coupling efficiency to the waveguide, which is less than ideal due to limited resolution of the MIR optics at 3.4-μm wavelength. About 4 μm of (theoretical) resolving capacity was determined for the best optics available in the lab, which is close to the mode-size of about 20 μm by 10 μm. Also, butt-coupling from the waveguide to the fiber is critical due to the different wavelengths involved. These problems could be alleviated by using highly symmetric waveguides with circular modes, for example using ridge waveguides or even photonic wires [48, 29]. Alternatively, the fiber optic coupling and filtering scheme could be replaced by a bulk-optic solution: in that case, coupling could be optimized for both the in-coupling of pump radiation, as well as the out-coupling of up-converted radiation, by using a beam combiner setup optimized for the two wavelengths. The filtering could be realized using a wavelength selective prism or grating arrangement, as used in the 1310 nm up-conversion spectrometer [59].

4. Absorption Spectroscopy using Frequency Conversion

Fundamental (rotational-) vibrational transitions of molecules, for example with hydrocarbon and hydroxyl bonds and respective compounds, can be excited by electromagnetic radiation in the 3-4 μm wavelength-range ([12] and references therein). The MIR wavelength-range in general is well-known as the fingerprint region, since resonant wavelengths are characteristic for any molecular species. While the absorption cross-sections are large compared to overtone-transitions in the NIR and visible ranges, better experimental conditions (e.g. due to more sensitive detectors) at these wavelengths have outweighed the stronger absorption in the MIR in industrial applications [60]. Consequently, the development of DFG-based MIR-sources and of up-conversion detectors can facilitate the extension of existing techniques to the MIR-range in order to exploit both specialized instrumentation, as well as strong absorption cross-sections.

There is considerable interest in detecting molecular compounds in gases, liquids and solids due to a number of applications where such constituents play a role [61]. The following list does not claim to be exhaustive, but is intended to give an overview of important applications where optical detection is useful:

- Combustion gas diagnostics [62]

- Trace gas monitoring (e.g. using satellite based spectrometers [63] or ground based safety installations [64])

- Medical analysis [31]

- Quality control of foods and pharmaceutical products [65, 66].

Investigation by optical spectroscopy has proven to offer many beneficial aspects. It can be fast [62], which is important when dynamics are investigated; specialized techniques like frequency-modulation spectroscopy can make it extremely sensitive [52]; and it is highly selective due to the fingerprint-like absorption response of different molecules, so the concentration of single constituents in a sample may be determined as well as the composition of a sample. In addition, it allows contact-, destruction-, and contamination-free investigations and facilitates sensing of remote samples in general.

There are numerous approaches to optical spectroscopy. In any case, information in the form of absorption at selected wavelengths is gathered using suitable detectors. Spectra can be measured using a broadband source (e.g. background radiation of the surface of the earth for remote sensing using satellites), in which case spectral filtering with dispersive elements (e.g. using a grating spectrometer) is needed for wavelength selectivity. In this case, both the inhomogeneous illuminating spectrum, for example from a blackbody, as well as residual effects must be accounted for during data processing.

On the other hand, tunable laser sources may be used to perform spectroscopy by tuning the laser wavelength over absorption features. In this case, wavelength selectivity is intrinsically given. The spectral resolution of the measurement is basically limited by the laser linewidth, which may be orders of magnitude smaller than that of a grating spectrometer [52]. Lasers also opened new methods of spectroscopy: for example, a number of nonlinear phenomena like Raman scattering may be exploited for spectroscopic purposes. Also, pulsed lasers with pulse durations down to the femto-second range can be used for time-resolved investigations at ultra-short time scales [67].

There have been some very exciting efforts to realize DFG-based MIR-spectrometers. For example, Weibring et al. demonstrated for the first time a high performance DFG spectrometer on airborne platforms, mixing a 1083-nm pump with a 1562-nm tunable signal source in bulk PPLN to generate 3.5-μm radiation [68].

In [12], I already demonstrated that Ti:PPLN waveguides, pumped with (tunable) diode lasers, are an excellent source for narrow linewidth, tunable MIR-radiation in the 3-μm wavelength range. Some results from that thesis are reviewed here to show the high quality spectra obtained using a Ti:PPLN-based DFG-MIR source. With pump powers in the milliwatt-range, and output powers in the microwatt-range, essentially noise-free measurements were performed with lock-in technique and small time constants.

Within the framework of this dissertation, the application of Ti:PPLN based DFG-MIR sources together with SFG-based up-conversion detection to the field of absorption spectroscopy is demonstrated. MIR radiation is used to access the strong, fundamental vibrational-rotational absorption bands of methane, with up to two orders of magnitude stronger absorption bands than in the NIR. Using the quasi-phasematched, Ti:PPLN based devices described in previous chapters, wavelength tuned MIR-spectroscopy of methane was conducted. In the following section 4.1, a short review of absorption spectroscopy is given. In the subsequent section 4.2, the foundations laid in [12] are reviewed, where DFG-MIR sources were used for MIR-generation and conventional MIR-detectors were used for detection. In section 4.3, the addition of a fiber-coupled, SFG-based up-conversion detector is demonstrated, realizing a MIR laser spectrometer which relies on NIR-instrumentation only.

4.1. Basics of Trace Gas Spectroscopy

In the most basic form of laser absorption spectroscopy, radiation at a specific frequency ν is transmitted through an absorbing sample under investigation, e.g. a gas cell of length L with an entrance and an exit window (as shown in the DFG-based spectrometer in figure 4.1).

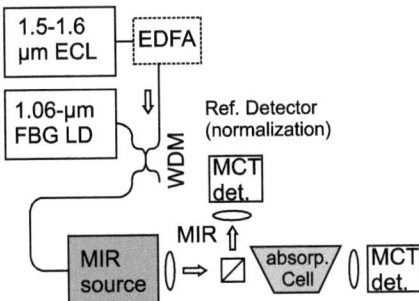

Figure 4.1.: Absorption spectroscopy setup using conventional detection. Pump and signal source for MIR-generation are the tunable ECL and the fixed-wavelength, 1064-nm FBG-stabilized laser diode. An EDFA can be used to boost signal power but was not used in absorption measurements. A beam splitter is used to reflect 50 % of MIR-radiation to a reference detector for power normalization. (Polarization controls omitted for clarity.)

The evolution of an initial intensity I_0 along a path L with a homogeneous gas composition is described by the law of Lambert-Beer according to

$$\frac{I(L)}{I_0} = \exp(-\gamma L), \qquad (4.1)$$

where $\gamma = \alpha_i + \tau$ is called the extinction coefficient, and L is the optical path length. $\alpha_i = k_i \rho_i$ describes the absorption which is to be determined, with absorption coefficient $k_i(\lambda, T)$ and molecular density ρ_i specific to a molecular species i. Due to broadening effects of absorption lines, the absorption coefficient $k_i(\lambda, T)$ can be a function of several absorption lines itself. The coefficient τ describes extinction from all other sources, e.g. due to scattering, cross-sensitivities, etc.

If the wavelength-dependence of $\tau(\lambda)$ is known, one method to determine α_i is to perform a differential absorption measurement at two selected wavelengths, so that

$$-\frac{1}{L}\left(\ln\frac{I_{\nu_1}}{I_{0,\nu_1}} - \ln\frac{I_{\nu_2}}{I_{0,\nu_2}}\right) = (\alpha_{\nu_1} + \tau_{\nu_1}) - (\alpha_{\nu_2} + \tau_{\nu_2}) \qquad (4.2)$$

Ideally, τ is wavelength independent, so that τ_{ν_1} and τ_{ν_2} cancel out in equation 4.2. In that case, we find the species mole fraction

$$\chi_i = k_B T \frac{\alpha_{\nu_1} - \alpha_{\nu_2}}{(k_{\nu_1} - k_{\nu_2}) P_{total}}, \qquad (4.3)$$

with the Boltzmann-constant k_B, temperature T, and total pressure P_{total} of the gas mixture under investigation[1]. The species mole fraction χ_i can now be determined, if

---

[1] The equation follows from the relation: $\alpha_i = k_i \rho_i = k_i P_i \frac{P_i}{k_B T}$, with partial pressure P_i of the absorbing species. The species mole fraction is given by $\chi_i = \frac{P_i}{P_{total}}$.

the differential absorption coefficient $k_{\nu_1} - k_{\nu_2}$ is known. The absorption coefficient is in general a complex function of wavelength, pressure, temperature and gas mixture. It may be obtained either from calculations (see e.g. [52]) using spectral databases like HITRAN [69], or from calibration measurements [62].

4.2. Absorption Spectroscopy using DFG-MIR Source

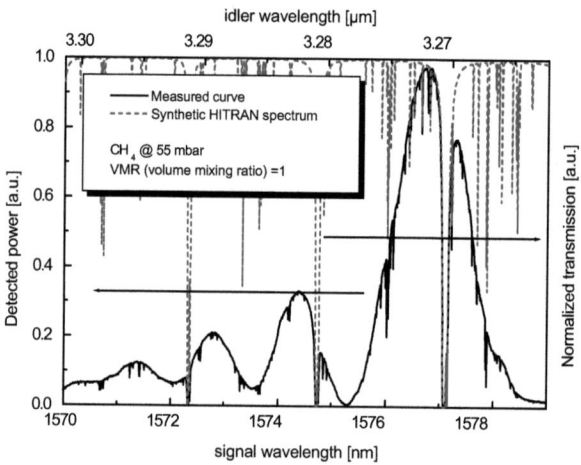

Figure 4.2.: Absorption measurement using DFG-MIR source and a MCT detector [12]. The ordinate shows the wavelength-dependent output power of the QPM device, with absorption lines from pure methane (VMR of 1) at 55 mbar in a 39.5-cm long gas cell (black line). A corresponding transmission spectrum was calculated as well, using the HITRAN database [69] (red line).

The feasibility of using Ti:PPLN waveguides for MIR-generation for absorption measurements was previously demonstrated in the Integrated Optics group Paderborn, using conventional MCT detectors [12]. Measurements can easily be normalized by splitting the incident beam by employing a beam splitter and a reference detector, as shown in figure 4.1.

An ECL operating in the 1550-nm wavelength range was used as wavelength-tunable signal source, with a fixed 1064-nm pump. In figure 4.2, the typical tuning characteristics of a waveguide sample is displayed, with absorption from pure methane at 55 mbar of pressure at room temperature. The gas cell was 39.5-cm long with CaF_2 Brewster-angle windows. A corresponding absorption calculation was performed using line data from the HITRAN database [69], considering pressure- as well as Doppler-broadening.

Further normalization of measured spectra emphasizes the outstanding agreement between measurement and calculation, as can be seen in figure 4.3.

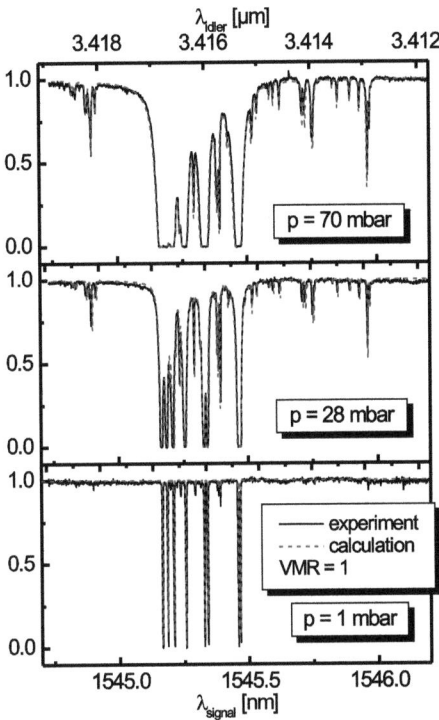

Figure 4.3.: Absorption measurement using the aforementioned setup. The reference detector was used for power normalization of the spectrum. Transmission is shown as function of methane content of the 39.5-cm long absorption cell, with a VMR of 1 (i.e. 100% methane). Nearly total extinction is reached even at 1 mbar of methane, at selected absorption lines [12].

For a practical application, the measurement could be restricted to a differential signal between two discrete wavelengths; one with an absorption feature, and another spectral position free from absorption. The mole fraction would be then determined according to equation 4.3.

4.3. Absorption Spectroscopy using DFG-MIR Source and SFG-UCD

The hybrid UCD now replaces the MCT detector [70] (figure 4.4). Here, signal radiation

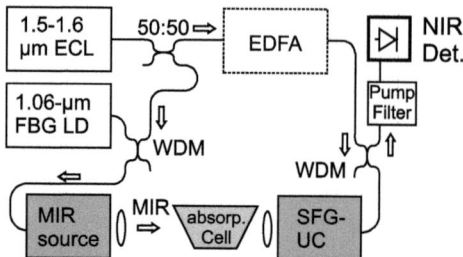

Figure 4.4.: Absorption spectroscopy setup using an SFG-based up-conversion detector. Radiation from the ECL is split using a 50:50 fiber-coupler. The second output from the coupler is used to pump the SFG-based up-converter (UC), which forms the UCD together with the NIR detector (an InGaAs detector was used in this case). Pump power may be amplified by an EDFA if desired. (Polarization controls omitted for clarity.)

is split using a fiber-based 50:50 coupler in order to distribute the power among the DFG-MIR and SFG-UCD modules. An EDFA may be used to boost the power used to pump the up-conversion detector, however it was not used for these measurements, due to sufficient conversion efficiency of the modules to measure clear absorption spectra in a laboratory environment. About 1.15 milliwatts external signal power were available for each module, whereas the MIR-source was pumped with the 32 milliwatt FBG-stabilized laser diode, giving about 4.6 μW of idler power. About 2 nW of up-converted power could then be detected with the InGaAs detector.

Both modules had been temperature-tuned to match phase-matching conditions (see chapter 5 for a more detailed discussion on transmission characteristics in a largely similar setup). With the setup shown in figure 4.4, spectroscopy in the range of about 3.2-μm to 3.6-μm wavelength may be performed, with an instantaneous tuning range of 3.6 nm in the MIR (restricted by the phase-matching bandwidths). In the photo in figure 4.5, the transmitter and receiver modules, as well as the absorption cell can be seen. In the background, both MIR camera and MCT detector are visible (not in use); the ECL used as signal source is also to be seen (the DFG pump source is blocked from view).

An example spectrum is shown in figure 4.6. Periodicity and temperature were chosen in such a way that the phase-matching wavelength coincided with a strong absorption feature of methane. Again, the phase-matching characteristics of both, transmitter and receiver modules, account for an envelope superposed with the pressure-broadened absorption lines. A single-pass gas cell of 28-cm length was used in this case. Normalization

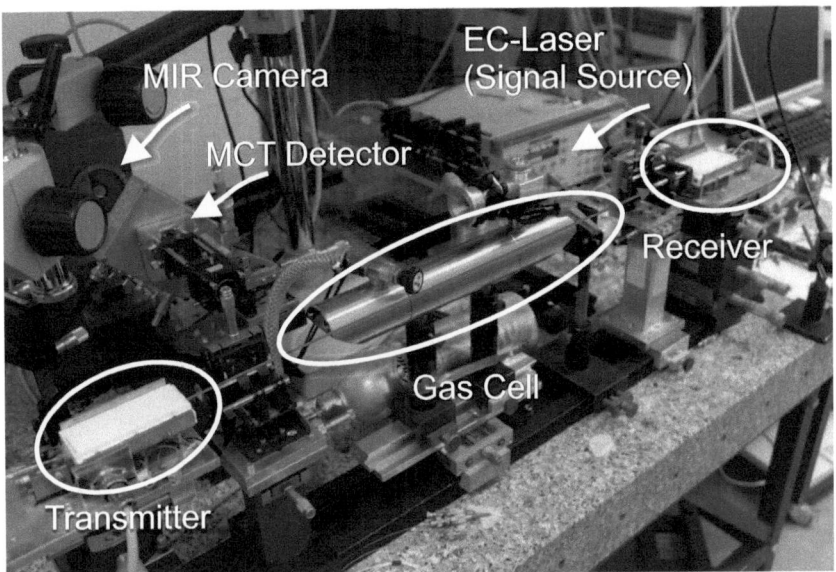

Figure 4.5.: Photo of absorption measurement setup using DFG-MIR generation and SFG-UCD. MIR radiation is coupled from transmitter- to receiver-waveguide (pump and signal radiation are blocked using a Ge-filter).

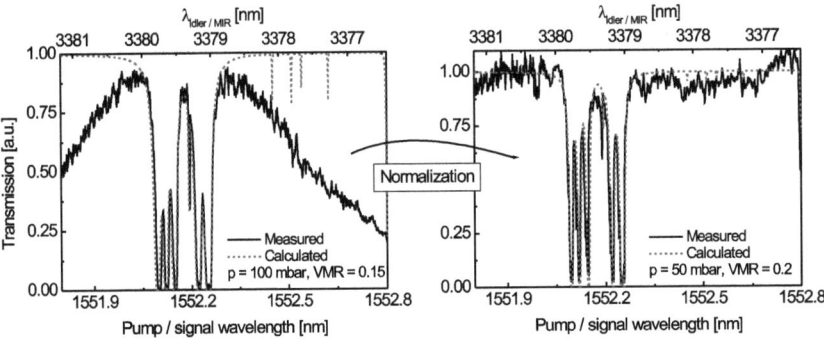

Figure 4.6.: Methane absorption measurement, using DFG-MIR generation and the SFG-UCD. Normalization was done by subsequently measuring the phase-matching curve with and without the absorption cell (note that different measurements are shown, however at the same spectral position).

was also realized, in this case by subsequently measuring the transmission characteristics with and without methane absorption. Due to slight instabilities in the setup, of mechanical nature as well as due to slight drifts of the sample temperatures, perfect normalization was not achievable. However, there is potential to improve stability and the normalization technique, so this is not a fundamental problem but a practical one.

The attractiveness of this concept is twofold. The flexibility of DFG can be exploited to tune the MIR radiation to a desired wavelength, e.g. to a specific absorption feature free of cross-sensitivities. On the receiver side, the exceptional sensitivity of the UCD can be fully exploited, using wavelength-matched, nonlinear modules. The spectrally selective properties of the UCD are nearly ideal for laser spectroscopy. In addition, fiber-optic technology can be employed, operating in the NIR. This could be potentially interesting considering a concept presented by Culshaw et al. [64], who developed a detection scheme exploiting the 1.67-μm absorption band of methane. Optical fibers were used in this case to distribute radiation to several measurement sites. Combining both, a fiber-optic network, and the DFG-MIR source and SFG-UCD, could improve sensitivity while retaining the flexibility of a centralized supervision scheme.

4.4. Frequency Modulation Spectroscopy using DFG Source and SFG-UCD

Advanced spectroscopic techniques might as well be used. In [12], for example frequency modulation spectroscopy was demonstrated using the DFG-MIR source with conventional detection using the MCT detector. The technique can be transferred easily to the setup previously described. In this case, the ECL was frequency modulated with a modulation depth of ca. 800 MHz (corresponding to ca. 60 pm in the MIR). The setup is drawn in figure 4.7.

Due to the modulation of the laser wavelength, the measured signal is proportional to the slope of the measured curve. Using this technique, the center of the absorption peaks can be easily made out at the zero crossing. Also, the technique is less susceptible to amplitude noise due to the kind of phase-sensitive detection used.

A measured spectrum is displayed in figure 4.8. A slight impact of the phase-matching characteristics is noticeable. A transmission spectrum of a DC measurement is superimposed with strong absorption features of methane. Evidently, the amplitude corresponds to the slope of the transmission measurement, and fine absorption features are easily located.

4.5. Summary

Within the scope of this thesis, a spectroscopic setup for molecular absorption spectroscopy in the 3-μm range was realized, based solely on NIR instrumentation (lasers, detectors, fiber optics). The results were achieved building on previous work done

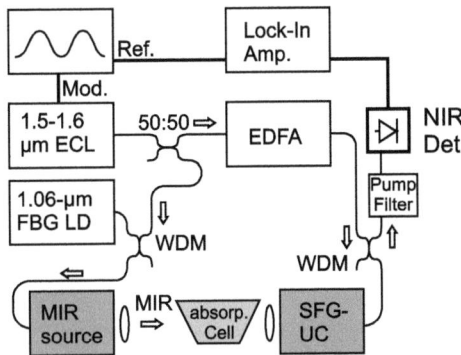

Figure 4.7.: Frequency (or wavelength) modulation spectroscopy setup using an SFG-based up-conversion detector. The ECL is frequency modulated by piezo-control, using a sine-wave from a function generator. The *modulation* frequency is used as input for the lock-in amplifier.

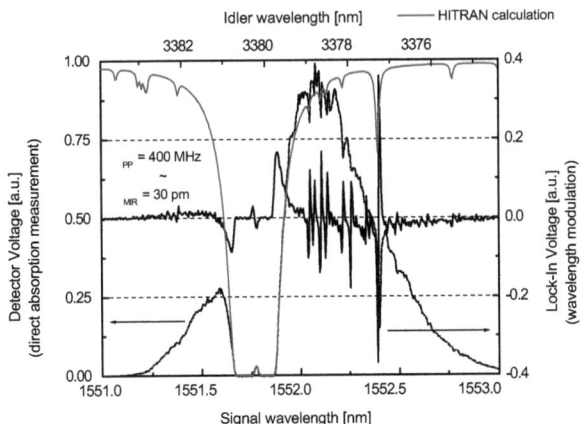

Figure 4.8.: Wavelength (or frequency) modulation spectrum of methane, using DFG source and SFG-UCD. The wavelength-modulated spectrum, with about 30-pm modulation depth in the MIR, is superimposed over the measured transmission spectrum (broken line). A calculated spectrum is also given. Pressure was 1 atm with a VMR of 2.5% of methane.

by Wiegand [13] and myself [12]. The concept of DFG based spectroscopy with up-conversion detection using waveguides was verified for the first time by Wiegand. Her results could be improved mainly by integrating the pump-filtering schemes for up-conversion detectors, which were presented in chapter 3, into the experimental setup developed by Wiegand. Investigation has shown that pump and signal powers in the milliwatt-range are sufficient as well for MIR-generation as for up-conversion detection, at least in the laboratory, in order to conduct MIR absorption spectroscopy. Powers can easily be increased by using a high power pump source and signal amplification using an EDFA to pump the DFG-MIR source (chapter 2). Also, an EDFA can be used in order to improve detection sensitivity of the SFG-UCD, up to what was presented in chapter 3.

The detection sensitivity was not determined, because it depends on a number of factors, such as the spectroscopic technique which is used, output power stability, etc. As the primary goal of this work was not to develop a highly sensitive trace-gas detection device, only little work was spent on developing a competitive spectroscopic setup. However, it may be estimated that the dynamic range of the InGaAs detector is about 85 dB at a gain setting of 10^4. According to the data sheet, 1 milliwatt of power would roughly saturate the detector in this case. This means that an absolute minimum modulation depth of $2.6 \cdot 10^{-9}$ could still be detected, assuming a linear detector response up to saturation. About 16 milliwatts of MIR power would be necessary to roughly saturate the detector, using the SFG-UCD with 1 Watt of pump. Theoretically, it would then be possible to measure methane concentrations much below 1 ppb / $Hz^{1/2}$ (ppb = parts per billion) with 1 meter of path length in an absorption cell, using the strongest absorption line of methane at 3260.2-nm wavelength. In practice, considerable effort is needed to achieve these figures. For example, radiation power should be highly stable and active stabilization methods might be necessary. Methane detection sensitivities in the range of tens of ppb / $Hz^{1/2}$ were reported previously, using DFG-based MIR spectroscopy [71].

The measurements using the up-conversion detector are somewhat more noisy, due to reduced mechanical tolerance of the setup - compared to the MCT-based DFG-spectrometer, for example - because of the sensitive coupling to the receiver waveguide and fibers. Also, the means to develop a scheme for near real-time normalization of the absorption measurements were not available. These are mainly practical issues which can be resolved in a stationary setup used for spectroscopy (e.g. using galvo steering mirrors, the beam path could be switched at high frequency to realize a reference beam path, in addition to the measurement path, in order to realize normalization). In turbulent environments (considering combustion diagnostics, for example), the small waveguide dimensions will invariably present an additional noise source compared to a larger area detector - although, as will be explained in the following and last chapter (also in appendix E), the optical effect of turbulence on MIR radiation is small compared to NIR radiation.

Frequency modulation spectroscopy could also be verified using the SFG up-conversion detector. Due to the zero-crossing at the center of an absorption line, this technique could for example be useful for absolute wavelength calibration of a DFG-based MIR laser, using a reference gas cell.

5. Free-space Optical Transmission in the MIR using Wavelength Conversion

We are living in the information age, where the better part of the world's population is gaining access to numerous streams of information, through radio, the telephone, television, and the Internet. The backbone of the Internet is formed by optical fiber networks, which exploit the tremendous bandwidth at optical carrier frequencies. Charles K. Kao was awarded one half Nobel prize in the year 2009 for his own, quote, "groundbreaking achievements concerning the transmission of light in fibers for optical communication" [72] during the 1960ies.

Figure 5.1.: Illustrated free-space optical links between land, air and space based systems.

Apart from fiber-optic networks used mainly for long-haul transmission of data, there are last-mile transmission systems using copper cables and electrical data transmission, or even glass optical or polymer optical fibers for optical data transmission. All these technologies rely on the installation of stationary cables. In the case that such installations transpire to be impractical or undesirable, the use of free-space optical (FSO) transmission systems is an alternative (radio-frequency (RF) transmission, like the IEEE 802.11 wireless local area network (WLAN) standard uses [73], is another). In principle, the large bandwidth of fiber-optical communication technology is retained when FSO transmission systems are employed, so transmission capacities with Gbps-bandwidth, and even WDM (Wavelength Division Multiplexing) systems are feasible for increased capacity or added redundancy. The collimated optical beam used in such a system, stemming from a coherent source with low beam divergence, features an intrinsically high directivity and eavesdropping resilience. From a technical point of view, a FSO system can easily be designed using standard telecommunications and fiber-optical components. Even quantum key distribution has been demonstrated using FSO transmission [74].

Possible applications for FSO systems are numerous. For example, individual buildings in on-campus or factory premises can be integrated into local area networks (LAN) using FSO transceivers. Applications with high requirements on mobility, for example in temporary disaster situations, are also thinkable. Obstacles like rivers or streets can easily be by-passed using roof-to-roof FSO links [75]. Even integrated space–terrestrial networks, with optical links between satellites, ground, and aircraft have been proposed and verified experimentally [76], which has been illustrated in figure 5.1.

In consequence, there is a rapidly growing market for free-space optical transmission systems, due to its flexibility and versatility. Most such systems operate in a wavelength range between the visible (>600 nm) and the communications band of fiber optics (<1600 nm) [77, 78]. The choice of wavelength depends on considerations of practicability, where economic and physical constraints should be addressed. For example, a suitable radiation source and corresponding optics must be available in the chosen wavelength range, as well as a modulation scheme with sufficiently high bandwidth for the design bit-rates. The system should be easily integrable into an existing, fiber-optic or wire-based network. Eye-safety requirements must be adhered strictly. In addition, wavelength- and distance-dependent impairments of the beam carrying the data bits need also be considered, with the application at hand in mind. As an example, clear-air turbulence as well as boundary layer turbulence will impact a link between a satellite and an aircraft [76].

It was found that, on the one hand, wavelengths in the MIR are favorable for certain FSO transmission regimes, and FSO transmission systems using QCL transmitters and MIR detectors were investigated. Aellen et al. used a 4.6-μm QCL together with an SFG-based UCD and a Si-APD detector as a feasibility study for Quantum Key Distribution (QKD) in the MIR [79]. They claimed that, in a calculated scenario with reduced visibility conditions due to fog, link performance is barely affected at 4.6-μm wavelength, whereas a system operating at 780-nm wavelength would simply fail to work.

On the other hand, coherent detection schemes offer potential benefits, because phase-distortion and fading effects due to atmospheric turbulence and multiple-access interference can be mitigated [76]. In that case, it is beneficial to be able to access integrated electro-optical technologies developed for the communications bands in the NIR, like heterodyne transmitters and receivers operating in the conventional optical C-band (1530-1565 nm) [80].

Within the framework of a project conducted in cooperation with the University of Stanford, CA, and CeLight, Inc., FSO data transmission experiments were performed using nonlinear optical wavelength conversion from 1550-nm wavelength to 3.8-μm wavelength, and back to 1550 nm. Free-space transmission of a Non-Return to Zero Quadrature Phase-Shift Keying (NRZ-QPSK) modulation scheme with a bitrate of 2.488-Gbit/s was demonstrated in the laboratory with a modest Bit-Error Rate (BER) penalty.

At around 3.8-μm wavelength, a transmission window exists in-between molecular absorption bands of CO_2 and water vapor. In addition, the harmful effects of scattering, as well as refractive index fluctuations in the atmosphere, become less pronounced with longer wavelengths (however, an actual performance estimation of such a system requires careful calculation, due to the accumulation of effects over the transmission distance [81]).

In the next section, the physical phenomena that hamper transmission are introduced. In appendix E, the theory is discussed in somewhat more detail and example transmission calculations using a model atmosphere are presented. Continuing in this chapter, characterization results of the nonlinear optical wavelength converters are presented; in analogy to the previous chapters 2 and 3. Experimental results, using these modules for FSO transmission through a wind tunnel simulating turbulent boundary layers of air at varying temperature, are discussed and references to additional published data are given.

5.1. Atmospheric Transmission Impairments

In order to take into account the multitude of effects the atmosphere exerts on radiation, or a coherent beam of light in particular, a number of theoretical models have been developed in the past to predict atmospheric transmission [82]. However, in order to accurately predict the transmission of radiation on a predefined path, a number of parameters must be known, for example the constituents of air, the density distribution of the air mass above the surface of our planet earth, the temperature distribution, etc. Standard atmospheres have thus been defined - for example the U.S. Standard Atmosphere [83], the ICAO (International Civil Aviation Organization) Standard Atmosphere [84], or the International Telecommunication Union ITU-R standard [85] - for the purpose of providing comparative models to communities like scientists and engineers.

Using the data provided, the transmission of a beam carrying data, for example originating from a satellite, traversing the different layers of the atmosphere, and impinging on a telescope at the surface of the earth, can be predicted. It is well known that optical transmission through the atmosphere is impaired by a number of phenomena, including [86]

- absorption: caused mainly by water vapor and trace gases, like carbon dioxide (CO_2);

- scattering: due to atmospheric particles, like dust or large molecules (smog), as well as smoke, mist, fog, rain, hail, and snow;

- scintillations and beam wandering: caused by temperature and thus refractive index fluctuations of air masses / turbulent eddies (optical turbulence);

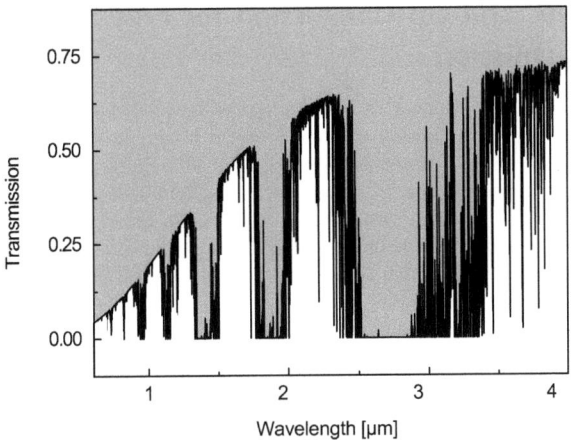

Figure 5.2.: Wavelength dependence of atmospheric attenuation, considering absorption as well as scattering, along a 2-km long horizontal path through clear atmosphere [87].

- as well as beam divergence / spreading due to finite apertures and aforementioned impairments.

In a realistic atmospheric transmission model, all of theses must be taken into account. Consequently, the wavelength of light should be chosen in such a way that the contribution of each to transmission impairment is kept low; i.e. bit error rates should be kept as low as possible and the availability of the FSO link should be maximized. Several transmission windows exist in the visible and MIR, where molecular absorption lines are absent. Especially scattering, however, leads to considerable attenuation of light at short wavelengths - simply consider the scattering especially of blue light in the atmosphere, leading to its familiar coloration. Also, scintillation by index fluctuations in air, which induces distortions of the wave fronts of an optical beam, have much less effect at longer wavelengths (albeit the accumulated effect must be considered, which has some additional implications; see appendix E). Beam divergence must also be taken into account, as well as background radiation of the earth - these are effects that restrict radiation to shorter wavelengths, approximately below 4 μm. A suitable transmission window is thus found at around 3.8 μm.

See appendix E for common mathematical treatments of each phenomenon, as well as exemplary transmission calculations. One straight-forward result is seen in figure 5.2. Atmospheric scattering causes strong attenuation at short wavelengths, whereas molecular absorption is distributed broadly due to the multitude of molecular species.

Of course, additional impairments, e.g. due to obstacles like mountains, buildings or even aircraft, may be present in reality.

5.2. Down- and Up-Conversion for Free-Space Optical Transmission

In the previous section, the use of MIR-radiation for FSO data transmission was motivated through the introduction of optical effects in the atmosphere. A wavelength-dependent investigation of which reveals a number of advantages in going to the MIR, although the effort is rather high due to technological challenges and increased beam divergence in case of an MIR transmission system. In principle, MIR lasers, namely QCL's, may be used for FSO transmission systems. They show potential for considerable modulation bandwidth with theoretical limits of several hundred GHz and output powers in the milliwatt range [88].

The advantage in going to a system which uses non-linear optical frequency conversion, however, is given by the facts that nearly arbitrary wavelengths can be accessed, and standard optical communication technology can be employed for modulation purposes. Also, a hybrid UCD as described in chapter 3 can be used for efficient detection and signal processing. For example, coherent detection can be used together with non-linear frequency conversion, as amplitude and phase information is preserved in the nonlinear wavelength conversion.

Consequently, hybrid transmitter and receiver modules were developed using Ti:PPLN waveguides and non-linear optical frequency conversion from 1.55-μm to 3.8-μm wavelength and back. As mentioned, LiNbO$_3$ is a suitable QPM material for wavelength conversion involving 3.8-μm radiation, and the thermal background from the earth is small at this wavelength. A 1100-nm pump laser was used in order to generate the MIR radiation through the DFG process, and to re-generate 1.55-μm photons in the receiver for detection. This scheme is depicted in figure 5.3.

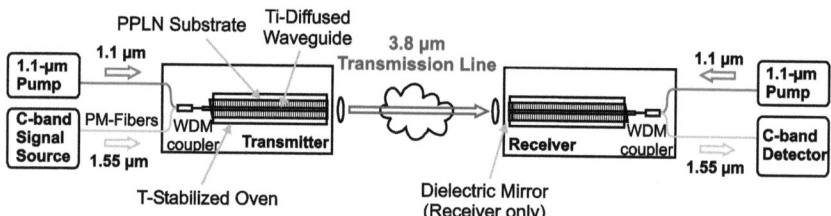

Figure 5.3.: Free-space optical transmission line in the mid-IR using frequency down- and up-conversion in Ti:PPLN waveguides. The transmitter and receiver modules are similar to the ones described in foregoing chapters, however polarization maintaining fiber components were used, and an 1100-nm pump to generate 3.8-μm radiation.

The input signal can be provided by any communication laser operating at 1550-nm wavelength. It is superimposed with the 1100-nm pump radiation using a polarization maintaining WDM coupler fabricated by AFW Technologies Pty. Ltd. As the modules

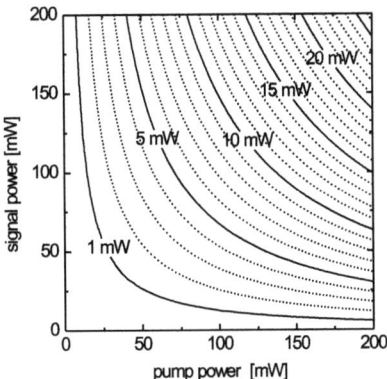

Figure 5.4.: Power characteristics of the FSO transmitter module, showing that about 25 milliwatts of idler power can be generated with 200 milliwatts of both pump and signal power. An 85-mm long waveguide was assumed for modeling.

operate on TM-polarized modes, the input signal must be linearly polarized in order for the wavelength conversion to function at the specified phase-matching wavelength.

Again, the waveguide internal power characteristics were calculated for the transmitter module, assuming an interaction length of 80 millimeters. The expected output power as function of pump and signal powers is shown in figure 5.4. The calculation shows that, using moderate pump and signal powers in the range of hundreds of milliwatts, tens of milliwatts of MIR power can be generated.

5.2.1. Experimental Setup

Before setting up a FSO transmission line, the transmitter and receiver modules were characterized individually. A common phase-matching wavelength was then selected by choosing appropriate waveguides and temperatures for each module.

The experimental setup used to characterize the individual modules / samples is similar to that described in chapter 2 (figure 5.5). The 1064-nm FBG-stabilized laser is replaced with a KOHERAS AMPLIPHOS 1100-nm fiber laser with an integrated amplifier. The output from the laser, in this case, is a linearly polarized, free-space beam collimated from a high-power single-mode fiber, so a fiber-coupling arrangement was used in order to couple the pump radiation to the respective WDM-port with a $f = 10$ mm lens.

Also, as polarization-maintaining (PM) PANDA-fibers were used in the communication experiments, the correct polarization for the pump beam was set using an AR-coated

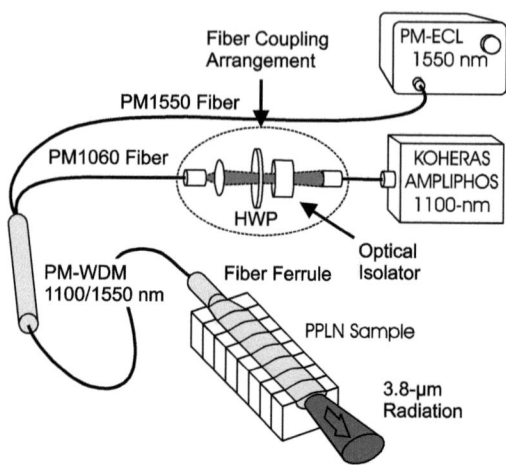

Figure 5.5.: DFG power characterization setup for FSO transmission modules, using polarization maintaining (PM) fibers. The linearly polarized free-space beam from the KOHERAS pump laser was aligned using a HWP, before coupling to the PM-fiber.

half wave plate (HWP) in the free-space beam, before coupling to the pm-fiber. A FSO-isolator was used in order to suppress excess back-reflections from the fiber in order to ensure stable pump operation. As signal source, a PM-ECL was used in this case; additional amplification could be provided by a PM-EDFA amplifier (omitted for clarity). Characterization results are given in the next section.

For the FSO transmission line characterization (figure 5.6), both modules were operated with the same pump, by splitting pump radiation using a polarization beam splitter (PBS) (this was done for convenience, because a single high-power pump was available; in principle it is not necessary to use the same pump for both modules). A signal at 1550 nm, coming from an ECL, was fed into the transmitter module. The signal was down-converted using the 1100-nm pump, generating 3.8-μm radiation coupled to free space using a f = 8.3 mm CaF_2 lens. The power distribution could be set by another HWP in front of the PBS; however, a 50:50 ratio proved to give the highest overall transmission through the system. In case of the receiver, a pump reflection mirror was deposited onto the sample end-face, similar to the mirror described in chapter 3. The mirror characteristics, measured on a witness-sample, are shown in figur 5.7.

5.2.2. Free-Space Transmission-Line Characteristics

The sample of the transmitter module was 83-mm long; the sample of the receiver was about 82-mm long. The samples were housed in an 80-mm-long, temperature stabi-

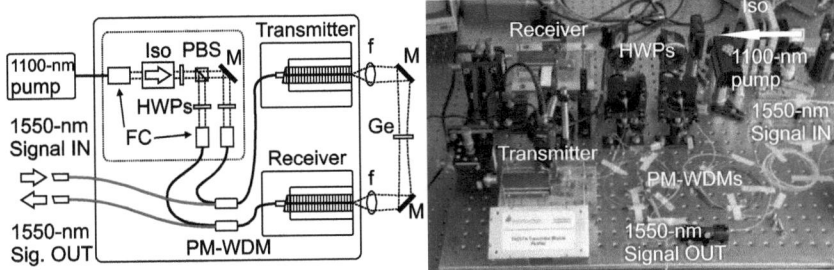

Figure 5.6.: Left: FSO transmission experiment using nonlinear wavelength conversion. Iso: Optical Isolator. PBS: Polarizing Beam Splitter. HWP: Half-Wave Plate. M: Mirror (Gold/Aluminum). Ge: Germanium Filter. f: f = 8.3 mm CaF_2 lens. FC: Fiber Coupling Optics. Right: Photo of the setup on an optical breadboard (mirror-symmetric to scheme).

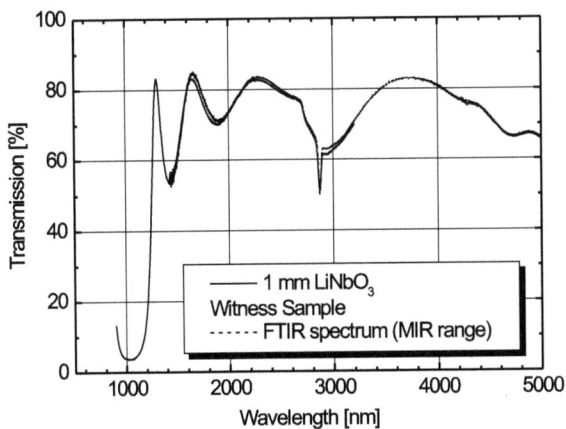

Figure 5.7.: Dielectric pump-reflection mirror of the receiver sample, which reflects pump radiation at 1100 nm, and transmits 3.4-μm radiation.

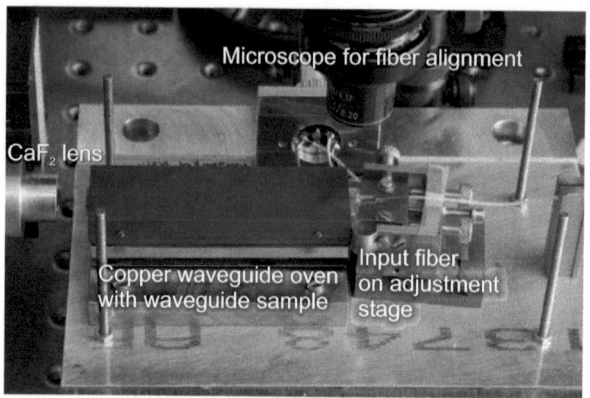

Figure 5.8.: Photo of the fiber-coupled transmitter sample in a temperature stabilized, heated Cu sample holder, operated at up to 200°C.

lized sample holder which could be heated to above 200°C in order to suppress harmful photo-refractive effects (figure 5.8). Pump and signal were combined by a polarization maintaining (PM) fiber-WDM coupler and butt-coupled to the waveguide using an 8° angle polished glass ferrule. The fibers were oriented to couple pump and signal to TM waveguide modes in order to exploit the strong d_{33} nonlinear coefficient of $LiNbO_3$. Due to the high operating temperatures, a micro-positioner was used to position the ferrule, instead of gluing the fiber to the sample. The out-coupled MIR-radiation was collimated by an f = 8.3 mm CaF_2 lens; a Ge filter was used to block residual pump and signal radiation in the free-space beam.

The tuning characteristics of both modules were measured operating both modules as transmitter. In figure 5.9, left, the normalized idler power is given versus signal wavelength at an operating temperature of 197°C. The maximum output power at this temperature is achieved at a signal wavelength of $\lambda_s = 1554.4$ nm, corresponding to an MIR wavelength of $\lambda_i = 3764$ nm. Again, a $sinc^2$-type wavelength response is expected in the ideal case. However, strong side lobes at the short wavelength side arise from waveguide inhomogeneities (variations in width and depth) as well as from temperature gradients towards the sample ends. These have been modeled by H. Herrmann [43] assuming a parabolic z-dependence of the phase mismatch term, $\Delta\beta(z)$, called 'chirp' (figure 5.9, right). The chirp corresponds to a variation of either temperature $\Delta T \approx 13°C$, Ti-strip width $\Delta w \approx 1.6$ μm, Ti-strip thickness $\Delta d \approx 5$ nm, or diffusion temperature $\Delta T_{diff} \approx 2.3°C$. Such inhomogeneities lead to a reduced efficiency at the peak wavelength as the enclosed area under the tuning curve is almost constant (The measured curves in figure 5.9 are normalized in this way). However, a potential transmission of data is unimpaired by side-lobes, as long as the signal bandwidth is significantly smaller than the phase-matching bandwidth. The phase-matching characteristics of both mod-

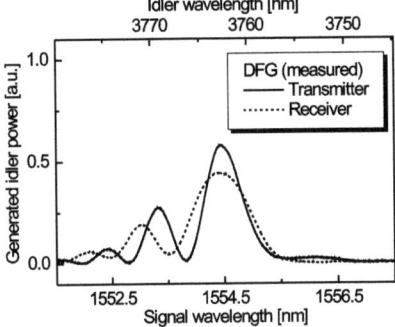 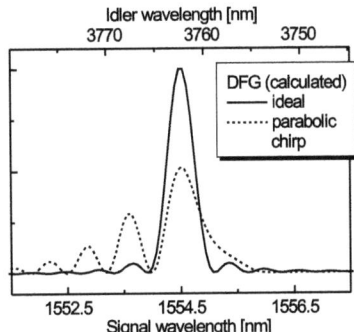

Figure 5.9.: Left: Measured wavelength tuning characteristics of Transmitter and Receiver modules. Right: Modeled tuning characteristics, assuming a parabolic chirp of the phase-mismatch term $\Delta\beta(z)$ [43]. The powers are normalized to a fixed area enclosed by the curves and the abscissa.

ules are very similar, but a shorter effective interaction length leads to a somewhat lower conversion efficiency and broader spectral response in the receiver module.

In figure 5.10, the measured MIR power from the transmitter module at $\lambda_s = 1554.4$ nm is plotted versus the product of coupled pump and signal powers at the input of the transmitter waveguide. These values were determined by power measurements in the free-space beam and subsequent correction for Fresnel losses, waveguide propagation losses, as well as residual losses in the Ge filter. In addition, calculated power characteristics from numerical simulations, assuming $d_{33} = 24$ pm/V, are shown. Surprisingly, the slopes of measured and calculated responses fit very well at small power levels although a smaller slope is expected for the experimental data due to the non-ideal measured tuning characteristics (figure 5.9). We attribute this discrepancy to uncertainties of some experimental parameters, of the nonlinear coefficient (in fact, it was probably overestimated by about 1/3 at the time of modeling) and of the waveguide model. At fixed signal power of 50 milliwatts, a linear increase of the idler power with pump power is observed with a pump-power normalized conversion efficiency of 69 %/W (I omit a more detailed treatment of uncertainties for clarity; see chapters 2 and 3 for some discussion). This characteristic should become super-linear at higher pump levels due to parametric amplification. At fixed pump power of 150 milliwatts, a roll-off is observed at high signal power levels, which is due to the onset of pump depletion. The maximum MIR power measured was 10.5 milliwatts when 150 milliwatts of pump and 130 milliwatts of signal power were coupled to the waveguide, corresponding to a power conversion efficiency of 8 %. The use of higher input pump power was forgone in order to prevent damage of the fiber optic WDM couplers. The measurement confirms that it is advantageous to use relatively high pump powers, compared to signal power, due to parametric gain.

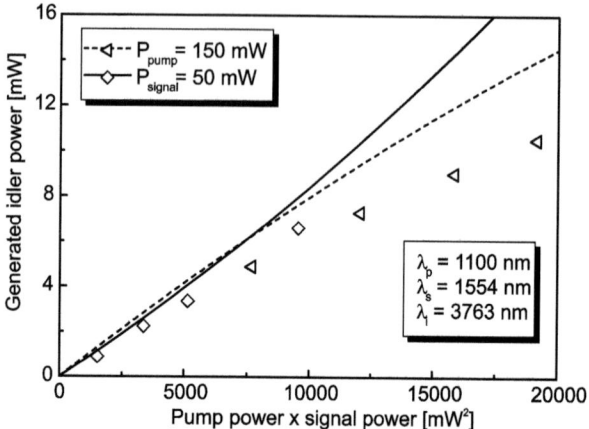

Figure 5.10.: Waveguide-internal idler power, generated by the transmitter module (deduced from measurements); depending on pump and signal power levels coupled to the waveguide. The lines represent corresponding calculations.

Using both modules, a 1.5-m long MIR-FSO link was set up to couple radiation from the transmitter to the receiver via two gold mirrors (see figure 5.6). Both modules were operated using a single pump laser followed by a 3-dB coupler. The wavelength dependence of the power regenerated by the receiver module is displayed in figure 5.11. The measured response agrees well with the product of transmitter and receiver tuning characteristics (expected response). At a MIR power level of 4.85 milliwatts in the FSO link, generated by pump and signal powers of 150 milliwatts and 80 milliwatts, respectively, coupled to the transmitter waveguide, about 100 μW (25 μW) of signal power was regenerated in the receiver waveguide (measured at the fiber output). Therefore, the transmission for the signal equals -29 dB, of which -14 dB can be directly attributed to the MIR-FSO link (Fresnel losses, residual losses in the Ge filter, collimation and coupling losses). The remaining -15 dB are due to the parametric processes in the two wavelength converters. Fiber optic, Fresnel and coupling losses from the signal source to the transmitter waveguide and equivalent losses from the receiver waveguide to the detector add additional -6 dB losses for each module, resulting in an overall fiber-to-fiber loss of -41 dB. The losses thus determined are shown in figure 5.12.

There is a large potential for reducing the losses of -14 dB in the MIR-FSO link. It is not quite clear why the free-space coupling losses are so high in this particular setup. However, I would estimate that a 10 dB improvement is feasible, using improved MIR lenses and more symmetric waveguides. The conversion losses of -15 dB due to the parametric processes can be reduced and even gain is theoretically possible by pumping the nonlinear waveguides with higher power levels. A pump power of 850 milliwatts

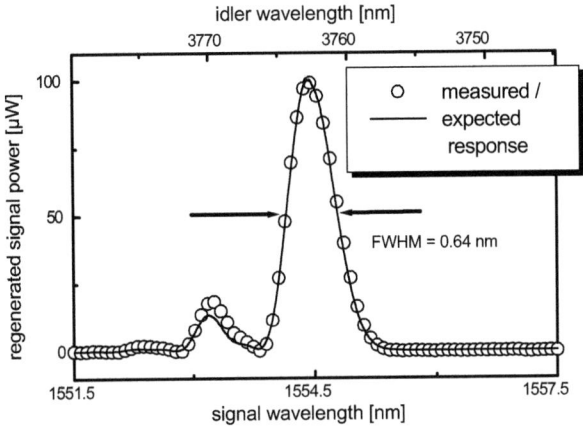

Figure 5.11.: Measured (circles) and expected (solid line) transmission line characteristics. The input signal power was 320 milliwatts. The expected line was generated by multiplying the tuning characteristics of transmitter and receiver modules (figure 5.9, left).

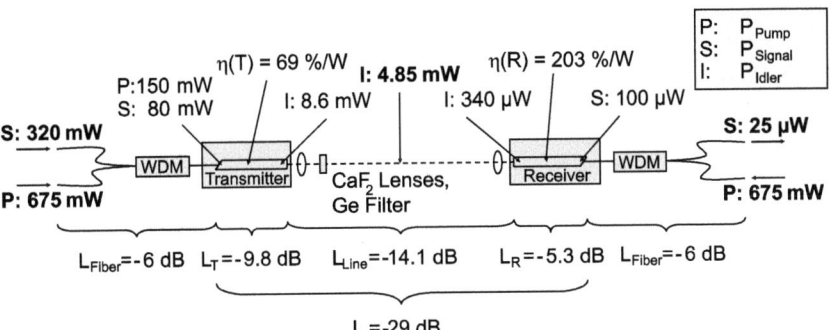

Figure 5.12.: Experimentally determined losses in the FSO transmission line. Bold power figures were measured; waveguide-internal power levels were determined by assuming plausible values for waveguide propagation losses (unknown at pump / signal wavelengths) and taking previously determined waveguide coupling coefficients; end-face transmission was taken from a $LiNbO_3$ witness sample coated with the end-face mirror dielectric stack.

coupled to both, the transmitter and receiver waveguides, could provide 15 dB improvement in conversion efficiency. However, this would require WDM couplers optimized for high-power operation; maybe even integrated on the LiNbO$_3$ substrate. An integrated pump-coupler on the substrate could also improve fiber-to-waveguide coupling (all in all, approximately -6 dB in each module), due to the separation of pump and signal radiation into two separate input waveguides. Again, symmetric waveguides, e.g. using ridge structures, could improve both mode-overlap as well as free-space and fiber coupling. In principle, an optimized lens-coupling arrangement to the fiber could also lead to increased power tolerance, and alleviate coupling losses.

All in all, if an improved fiber-coupling scheme and improved MIR optics were available, and sufficient pump power were coupled to the waveguides, the overall loss could be improved by an estimated 30 dB. In that case, the transmission losses would be reduced to -12 dB.

5.3. Evaluation of Free-Space Data-Transmission

Data transmission experiments via the MIR-FSO link were performed using the experimental setup described in the foregoing sections, and results were published by Ip [89]. The analogue quadrature phase shift keyed (QPSK) modulation setup used to conduct bit-error rate (BER) measurements at a bit rate of 2.488 GBit/s is shown in figure 5.13. A back-to-back measurement was done prior to determining BER with the two wavelength converters, and a BER impairment of 1.5 dB was observed by including the modules (figure 5.14 (a)).

Back-to-back performance was 14 dB below the additive white noise limit (figure 5.14 (a)). This was attributed to

- Quantum efficiency of the balanced photo-receiver (3 dB)
- Thermal noise due to insufficient local oscillator power (5 dB)
- Excess bandwidth of receiver low-pass filter (3 dB)
- Other losses (3 dB)

The 1.5-dB impairment caused by the wavelength converters was mainly attributed to amplified spontaneous parametric fluorescence. At 150 milliwatts of pump power at 1.55-μm wavelength coupled to the waveguide, parametric fluorescence is of the order of 1 nW within the phase-matching bandwidth of a Ti:PPLN waveguide of comparable length ([11], cp. also section 3.3.2, where comparable power levels where seen with a 1064-nm pump of similar power).

In addition to the demonstrated, one-directional transmission, receiver modules could also be operated as transceivers, enabling bidirectional communication. The transmitted and received wavelengths should however be slightly different within the phase-matching bandwidth, in order to avoid crosstalk due to the nonlinear process.

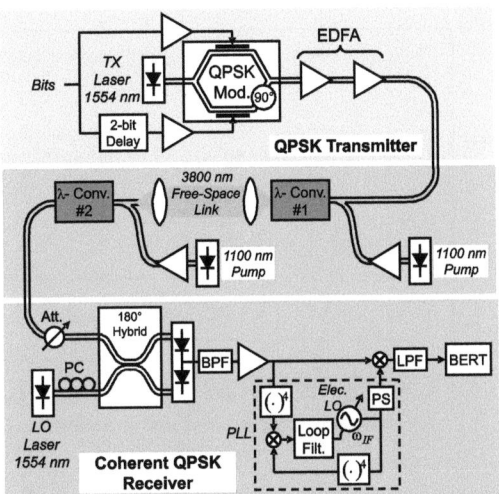

Figure 5.13.: FSO data transmission and bit-error rate measurement setup (courtesy of E. Ip [89]). The transmitter (here: λ-converter #1) is fed by a QPSK-modulated signal from a 1554-nm laser (TX, upper box). The signal is amplified by EDFA-stages. The signal is wavelength converted by the 1100-nm pump. The receiver (here: λ-converter #2) re-converts the 3.8-μm radiation back to a 1554-nm signal (middle box), which is mixed with a local oscillator (LO) in a 180° hybrid to generate two amplitude-modulated signals (sum/difference) which contain the original information from the QPSK modulator (lower box). The information is interpreted by signal processing electronics, and a Bit-Error Rate Tester (BERT) is used to evaluate transmission impairments.

The transmitter and receiver modules were as well used with a digital return to zero QPSK (RZ-QPSK) system by CeLight, Inc., to investigate the transmission performance of the modules, both with and without wavelength conversion to 3.8-μm, through a table-top wind tunnel. The basic setup is shown in figure 5.15. The RZ-QPSK system, with its own transmitting and receiving components, as well as BER-testing equipment, was simply connected to the input and output fibers of the wavelength converters for the investigations. Off-axis paraboloid mirrors with Au-coatings for the MIR with f = 2" were used in order to couple radiation into and out of the waveguides. The beam was then steered through CaF_2 entrance and exit windows of the table-top wind tunnel, using an arrangement of mirrors. In the wind tunnel, a number of heated rods would heat and perturb the airflow coming from a fan. Both rod temperature and airflow were varied in order to create turbulent conditions of varying intensity, with heating rod temperatures between about 100 and 250°C. A HeNe laser with 633-nm wavelength was

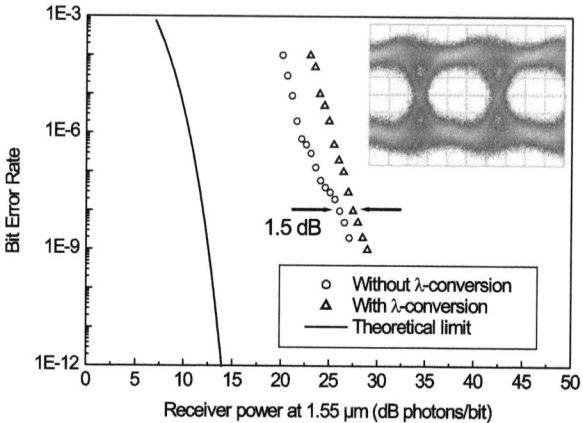

Figure 5.14.: Bit-Error rate (BER) measurement with / without wavelength converters. An impairment of approximately 1.5 dB is caused by the wavelength converters, which is attributed to amplified spontaneous emission. Inset: Opened eye pattern of the signal with wavelength converters, from a storage oscilloscope. A clearly opened eye is an indication for good signal quality. (courtesy of E. Ip. [89])

used in order to record a scintillation index, using a Shack-Hartmann sensor, in parallel to the transmission measurements - this could later be compared to the scintillation indexes at 3.8 and 1.55 μm, which were derived from data transmission measurements.

Data transmission using 3.8-μm wavelength and using 1.55 μm was compared at a bitrate of 160 Mbit/s of the RZ-QPSK signal (in the latter case of 1.55-μm transmission, the wavelength conversion modules were removed from the setup). In these measurements, absorption and scattering effects were negligible due to the short optical paths involved in the experiment, so scintillation provided the dominant optical impairment.

Under the most severe turbulence condition produced in the table-top wind-tunnel, 3.8-μm transmission would yield an estimated receiver sensitivity improvement of at least 6 dB over 1550-nm transmission. In the addendum to their paper done by associates of The Boeing Company, they discussed the theoretical wavelength-dependence of the scintillation index. To summarize the results, the scintillation index peaks at a certain propagation distance, after which it decreases and saturates. As the peak shifts to shorter distances with longer wavelengths, the use of 3.8-μm radiation is advantageous mainly for short distances. At the strongest turbulence conditions assumed, the advantage of 3.8-μm radiation is lost only after about 500-m propagation distance. Under less turbulent conditions, this turning point moves to longer distances. I have reproduced some of the theoretical considerations in appendix E.

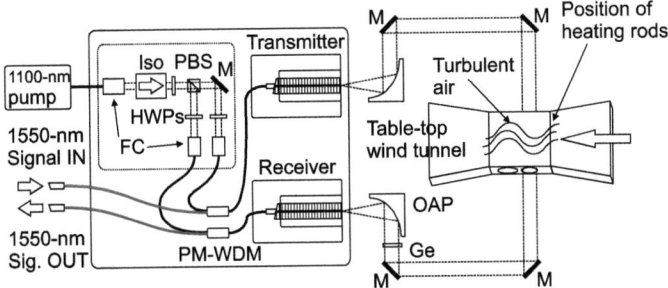

Figure 5.15.: Free-space transmission line through a table-top wind tunnel, using transmitter and receiver modules, in order to evaluate transmission performance at 3.8-μm wavelength compared to transmission at 1.55 μm. A red laser at 633 nm was used in parallel in order to record the scintillation index (not shown). OAP: Off-Axis Paraboloid mirror, Au-coated.

5.4. Summary

All-optical wavelength conversion is being intensively studied in the communications-wavelength range around 1550 nm [90, 91] for applications in future WDM-based transmission systems, using different approaches including semiconductor optical amplifiers, and nonlinear wavelength conversion using χ^2-processes. Moreover, χ^2-based DFG has been extensively investigated to generate MIR-radiation (see [92] for a nice overview), especially for spectroscopic applications. We could show that the same principles of χ^2 based wavelength conversion between NIR and MIR wavelengths can also be used for MIR-transmission systems based on established communications technology, and using frequency conversion as an interface to the free-space transmission link [43, 89].

Using the wavelength conversion modules developed in Paderborn, Ip et al. showed that free-space optical transmission using the wavelength converters cause a moderate bit-error rate impairment of 1.5 dB, compared to back-to-back performance of the system without free-space transmission.

Following these initial experiments, Cho et al. showed that, comparing MIR-transmission at 3.8-μm with NIR-transmission at 1550-nm wavelength, there is a considerable reduction of 6 dB in transmission impairments when transmitting through weak-turbulence gaseous boundary layers [81]. They compared transmission through a table-top wind tunnel, using the modules from Paderborn, with and without wavelength conversion. The wind tunnel simulated a turbulent boundary layer, provided by a number of heating rods disturbing the generated air-flow. Turbulent boundary layers are for example present above the surfaces of an aircraft, so the use of MIR radiation is expected to be beneficial for FSO laser-communications with aircraft.

Conclusions

Despite the advent of QCL lasers to the commercial market of mid-IR technologies, gaps still exist in the wavelength spectrum between the near- and mid-IR, especially when it comes to cw-operation and tunability. The long-standing lead-compound based (PbS/PbSe) lasers have some intrinsic difficulties, such as strong mode-hopping behavior and generally the need for cryogenic cooling, and precise control of operating parameters in order to select an operating wavelength.

Consequently, wavelength conversion by DFG has been and still is an attractive way to generate tunable radiation with wavelengths which are difficult to access by conventional means, especially in the mid-IR range. For the purpose of mid-IR generation, stable, room-temperature operated laser diodes with excellent spectral properties and considerable output powers, which are readily available in the near-IR, can be used. Within the scope of this work, waveguide-based wavelength converters, using Ti-indiffused waveguides in PPLN were thoroughly investigated. Lasers of 1064-nm and 1100-nm wavelength were used as pump sources, together with tunable lasers operating in the 1500- to 1600-nm range, in order to generate difference-frequency radiation in the 3.4- and 3.8-μm wavelength ranges. The results were presented in chapter 2, and MIR-output powers in excess of 10 milliwatts were achieved, with hundreds of milliwatts in the NIR. The tuning characteristics allowed instantaneous wavelength tuning ranges of about 5 nm in the MIR. In addition, temperature tuning of PPLN samples allows access to a broader wavelength range with a given waveguide sample - by tailoring the waveguide properties, nearly arbitrary wavelengths may be reached.

The concept of nonlinear wavelength conversion can also be used to up-convert MIR radiation for detection in the NIR. This has also been done in waveguide-based PPLN samples; the results were presented in chapter 3. The motivation is two-fold: on the one hand, MIR detectors are background-radiation limited, due to their sensitivity to thermal radiation, which NIR (InGaAs) or even visible light detectors (Si) are not. On the other hand, detectors with high modulation bandwidths are available especially for wavelength bands used in the optical communication technologies, which are optimized for optical fiber transmission. Two waveguide-based up-conversion detectors were investigated, using both SFG and DFG. In direct comparison to a TE-cooled HgCdTe detector optimized for 3.4-μm radiation, both SFG- and DFG-based up-conversion detectors outperformed the conventional detector by factors of 48 and 23.3. It is, however, difficult to compare the performance of detectors operating in different wavelength regimes quantitatively, due to the different causes that fundamentally limit NIR and MIR detectors. If, for example, the different detector sizes are taken into account (treating the UCDs as background-limited detectors), the advantage is reduced to factors of 4.2 and 2, respectively. In case of the DFG-based up-conversion detector, parametric gain is available, meaning that conversion efficiency is only limited by available pump power. However, parametric fluorescence limits the usefulness of this detector, because the fluorescence is present at the up-converted wavelength. In case of the SFG-based up-conversion detector, this is not the case; however, conversion efficiency is limited to 100% photon-conversion.

Both, transmitter modules using DFG for MIR-generation, and receiver modules using SFG for up-conversion, were used for molecular absorption spectroscopy of methane around 3.4-μm wavelength. This was reported in chapter 4. In these experiments, low NIR power levels in the milliwatt range, used to pump both the DFG-MIR source and the SFG-UCD, proved sufficient to conduct absorption measurements on methane with exceptional signal-to-noise ratios. It was also shown that frequency modulation spectroscopy, which is sensitive to the derivative of the absorption lines of a spectrum, is easily transferred to this particular spectroscopic setup. In direct comparison to absorption measurements using a conventional detector, the need to couple radiation to the SFG-UCD waveguide poses an additional (mechanical) noise source, which limits the possible detection sensitivity of the setup. The ultimate sensitivity was not determined, but the theoretical limit of conventional absorption spectroscopy of methane around 3.6-μm was estimated much below 1-ppb sensitivity, albeit experimentally, much effort is needed to reach values close to this.

Similarly, transmitter and receiver modules were used for free-space optical transmission in the MIR at 3.8-μm wavelength, which was presented in chapter 5. Calculations show that it can be beneficial to go to this wavelength, because scattering losses and scintillation effects are strongly reduced compared to radiation in the NIR or even visible, and transmission windows free of absorption lines also exist in this wavelength range. In addition, background radiation contributions from the earth is still fairly low below 4-μm wavelength. The modules developed in Paderborn were used to set up a free-space transmission line in the MIR at Stanford University, CA, using wavelength down- and up-conversion from the NIR to MIR and back. Transmission measurements were done together with Carsten Langrock of Prof. Fejer's and and Ezra Ip of Prof. Kahn's research groups, using an analogue QPSK system, and a data-transmission impairment of -1.5 dB only was caused by the free-space transmission line using the wavelength conversion modules. Later, the wavelength converters were moved to a Boeing facility near Seattle, WA. Here, the free-space transmission line was perturbed by a table-top wind tunnel, which simulated a turbulent atmosphere. Cho et al. could show, with a digital QPSK system connected to this setup, that frequency conversion to 3.8-μm using the waveguide modules provided a 6-dB gain in data transmission performance, compared to transmission at 1.55-μm.

In conclusion, it was shown that nonlinear-optical wavelength conversion to the MIR is a versatile means to access wavelength ranges which are especially interesting for spectroscopy and free-space optical communications. It is particularly intriguing that fiber-based optical components, for example from telecommunication technology, can be used directly to realize solutions for MIR applications. Ti-indiffused waveguides in PPLN offer a mature technological base for this purpose. Novel waveguide technologies with more symmetric optical modes with higher conversion efficiencies and better coupling to optical fibers, like ridge waveguides or even photonic wires, could improve the performance of such devices in the near future.

Appendices

A. Waveguide and Poling Mask Design

The photolithography masks for waveguide fabrication and periodic poling provided eight similar waveguide groups, and six poling sections with varied poling periodicities. The waveguides were bundled into groups of six, with two waveguides each of 18 µm, 20 µm and 22 µm waveguide width. The waveguide structure is shown in figure A.1.

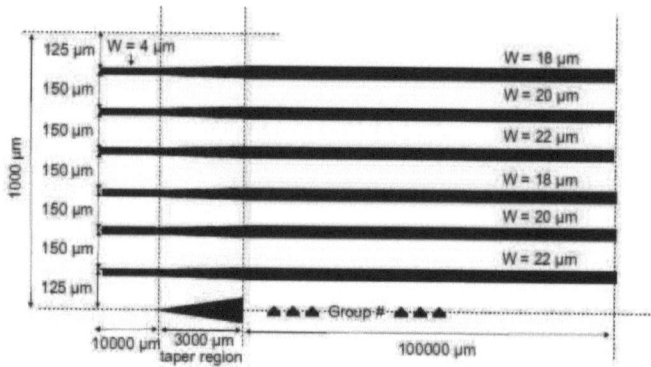

Figure A.1.: Layout of a single waveguide group with six waveguides. On the left side, the 4-µm wide input waveguide is shown, along with a tapered section used for selective mode excitation.

This design provides some redundancy (as in each poled group, two waveguides are available of each width), and also accounts for deficiencies in the waveguide model by allowing to select between different widths, in order to find the waveguides with optimum conversion efficiency.

Experimentally, the most narrow waveguides with 18 µm width generally showed the highest conversion efficiency. In theory, the highest conversion efficiency is expected for the 20-µm wide waveguides, at 170 nm Ti-layer thickness. The analytical waveguide model predicts improved conversion efficiency with increasing Ti-layer thickness (figure A.2); however, the amount of indiffused Ti is limited in practice.

Poling periodicities ranged from 26.5 µm to 27.25 µm - this way, the complete signal wavelength band of the Tunics ECL lasers in the range between 1500 and 1600-nm wavelength can be accessed, by selecting the corresponding periodicity and temperature for fine tuning (cp. figure 2.12). The mask layout is sketched in figure A.3. The poling

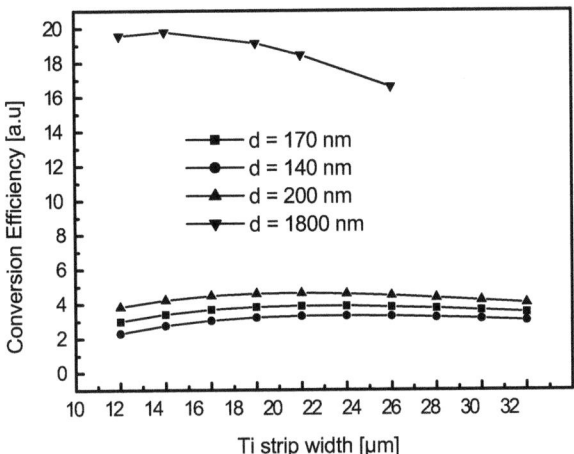

Figure A.2.: Nonlinear conversion efficiency (DFG) as a function of strip width and thickness. In the analytical model, conversion efficiency increases steadily with Ti-layer thickness. In reality, the amount of indiffused Ti is limited however when fabricating a functional waveguide. At 170 nm strip thickness, maximum conversion efficiency is expected between 18 and 22 μm strip width. Experimentally, the 18-μm wide waveguides generally showed the highest conversion efficiency.

mask was usually aligned to the central six waveguide groups of a single sample (marks were provided in the mask design for alignment).

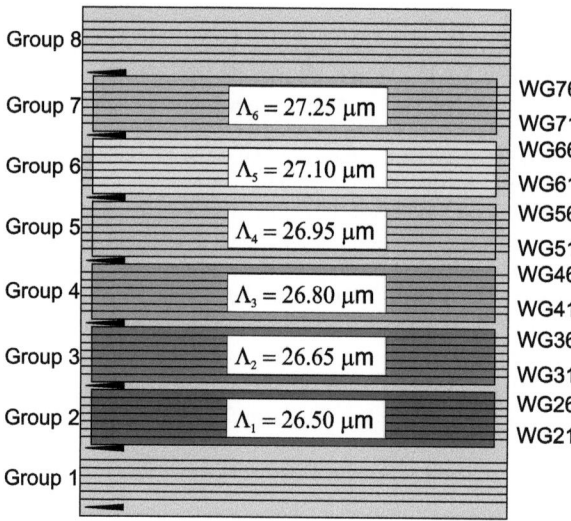

Figure A.3.: Mask layout with six poling groups. Poling periodicity Λ ranges from 26.5 μm to 27.25 μm.

B. Modeling of Lens Dispersion

The lens dispersion effect described in section 2.2 was modeled using Gaussian beam optics. The complex beam parameter of a beam propagating in z-direction is given by

$$\frac{1}{q} = \frac{1}{R} - i\frac{\lambda}{\pi w^2(z)},$$

with phase front curvature R, wavelength λ and beam half width $w(z)$. The beam half width can be determined from the imaginary part of $\frac{1}{q}$. A matrix formalism may then be used to determine the evolution of q, for example as function of propagation distance, or as it is modified by an optical component like a lens, as follows:

$$q_2 = \frac{Aq_1 + B}{Cq_1 + D}$$

with the matrix

$$M = \begin{pmatrix} A & B \\ C & D \end{pmatrix}.$$

For a given optical system, the matrices for each individual optical component or propagation path can be multiplied according to the laws of linear algebra in order to determine q at any point in the system. From that, w can be determined. This has been done for a simple arrangement of two BK7 lenses (f_1, f_2) placed 20 cm apart, both with f = 10 mm at 1550-nm wavelength. For the initial beam waist, the mode sizes of 1060XP fiber [93] at 1064-nm and 1550-nm were assumed. Modeling shows that the focal planes lie 0.1 mm apart. The Rayleigh length of the beam at 1064-nm was determined to be just 28-μm, in comparison.

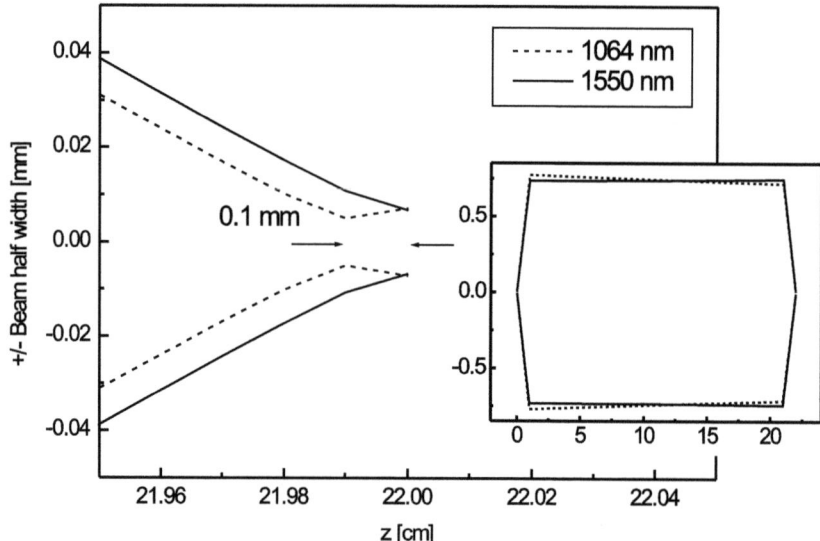

Figure B.1.: Gaussian beam modeling of two beams coming from a fiber. Two lenses (BK7) with f = 10 mm were assumed at z = 1 cm and z = 21 cm. At z = 0 cm, the output from a single-mode fiber was assumed. At z = 22 cm, the focal spots at 1064-nm and 1550-nm are about 0.1 mm apart. Inset: Beam profile along the 22 cm transmission path.

C. Waveguide Inhomogeneity Calculations

Asymmetries in the phase-matching tuning characteristics are caused by inhomogeneities of the Ti-indiffused waveguide. The most likely causes were already listed in chapter 2 and are:

- variations in Ti strip width w_{Ti},
- variations in Ti strip thickness d_{Ti},
- variations in the poling periodicity Λ,
- temperature gradients in the indiffusion oven T_{diff}, and
- temperature gradients of the sample during operation T_{op}.

In the coupled mode equations, the phase-mismatch term becomes $\Delta\beta = \Delta\beta(z)$, and the phase factor $\exp(\pm i\Delta\beta L)$ is generalized to the integral form:

$$\exp\left[\pm i \int_0^L \Delta\beta(z)dz\right] \approx \exp\left[\pm i \left(\Delta\beta_0 L + \int_0^L \frac{\Delta\beta(z)}{du}\Delta u(z)dz\right)\right],$$

assuming a deviation from the perfectly phase-matched interaction. The parameter u could be any of the above (w_{Ti}, d_{Ti}, Λ, T_{diff} and T_{op}).

In order to model the measured curves, a parabolic variation, centered about the waveguide center, was assumed for the argument $\frac{\Delta\beta}{du}\Delta u(z)$ (the model can, of course, be extended to more complex functions). The amplitude was varied in such a way that the resulting curve resembled the measured one closely; resulting in figure C.1. The parabolic chirp was then traced back to the parameters listed above. This was done by calculating the partial differentials

$$\frac{\partial}{\partial u}\Delta\beta = 2\pi\left(\frac{1}{\lambda_p}\frac{\partial}{\partial u}n_{eff}(\lambda_p, u) - \frac{1}{\lambda_s}\frac{\partial}{\partial u}n_{eff}(\lambda_s, u) - \frac{1}{\lambda_i}\frac{\partial}{\partial u}n_{eff}(\lambda_i, u) - \frac{\partial}{\partial u}\frac{1}{\Lambda(u)}\right).$$

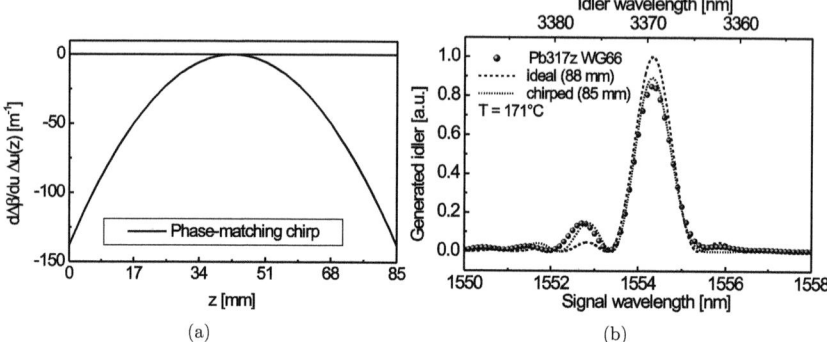

Figure C.1.: (a) Phase-matching chirp, assumed to reproduce the measured tuning characteristics. (b) Corresponding measured and calculated tuning characteristics (see also chapter 2).

The results are as follows:

$$\partial_{w_{Ti}}\Delta\beta = 229.8\,\frac{\mathrm{m}^{-1}}{\mu\mathrm{m}}$$

$$\partial_{d_{Ti}}\Delta\beta = 80.34\,\frac{\mathrm{m}^{-1}}{\mathrm{nm}}$$

$$\partial_{\Lambda}\Delta\beta = 8.619\,\frac{\mathrm{m}^{-1}}{\mathrm{nm}}$$

$$\partial_{T_{diff}}\Delta\beta = -162.7\,\frac{\mathrm{m}^{-1}}{^{\circ}\mathrm{C}}$$

$$\partial_{T_{op}}\Delta\beta = 18.7\,\frac{\mathrm{m}^{-1}}{^{\circ}\mathrm{C}},$$

using the mathematical convention $\partial_u = \frac{\partial}{\partial u}$. These results yield the maximum variations of

$$\partial w_{Ti} = -0.6\,\mu\mathrm{m}$$
$$\partial d_{Ti} = -1.72\,\mathrm{nm}$$
$$\partial\Lambda = -0.016\,\mu\mathrm{m}$$
$$\partial T_{diff} = 0.85^{\circ}\mathrm{C}$$
$$\partial T_{op} = -7.4^{\circ}\mathrm{C},$$

already presented in table 2.1, which are close to what was determined experimentally. Similar considerations were taken in order to model the measured tuning characteristics in chapter 5.

D. Reference HgCdTe Detector Characterization

In order to evaluate the UCD performance, an HgCdTe (or MCT) detector unit from OEC GmbH [94] was thoroughly characterized as the 'benchmark'-MIR detector.
The detector consists of a photo-conductive detector chip which is temperature stabilized to 213 K using a three-stage thermo-electric cooler. The chip is mounted in a hermetically sealed housing with an AR-coated window optimized for the 3.4-μm wavelength range, while the actual responsivity of the chip extends from the visible to 4.5-μm wavelength. The detector chip is integrated into a voltage divider circuit; the voltage drop across the chip is amplified by a low-noise voltage amplifier with 1-kHz respectively 10-kHz bandwidth (switchable), and a lower threshold frequency of 3 Hz (optionally DC). The integrated amplifier gives a DC-coupled amplification factor of 668, leading to detector saturation at a few tens of microwatts incident power.

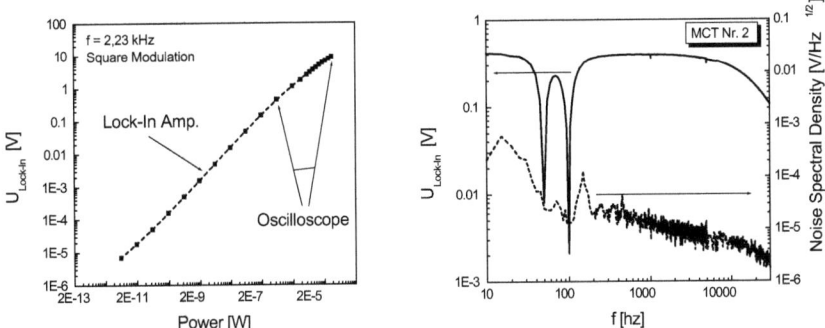

Figure D.1.: Left: MCT detector response curve, right: MCT spectral noise characterization.

Firstly, the power characteristics were measured and secondly, noise performance was evaluated using the DSP lock-in amplifier. The result is shown in figure D.1. In figure D.1, left, the detector response is shown at different power levels incident onto the detector. A distinctive non-linear behavior is evident at the lowest and highest power levels measured, mainly due to internal voltage amplification in the detector unit. In figure D.1, right, a frequency dependent measurement of the noise spectral density S_N (with 'dark' detector) is displayed, using a lock-in noise bandwidth of 17.2 Hz (time constant: 10 ms) [95], alongside the frequency response of the measurement system at a fixed incident power level. Beforehand, the sensitivity limit of the lock-in itself was found to be $S_N = 4.9$ nV/$\sqrt{\text{Hz}}$, which is about three orders of magnitude below the noise level measured using the detector.

Some interesting characteristics can be recognized from this measurement. The two dips in voltage at 50 and 100 Hz are evident, due to the line rejection filter in the lock-in amplifier to filter out electrical frequency components at the line frequency and its second harmonic (to filter e.g. ambient light). It is evident that the noise spectral density reduces with higher frequencies. However, the detector response also decreases due to bandwidth restrictions imposed by the detector circuitry. The optimum signal-to-noise ratio can thus be found within the vicinity of 1-kHz modulation frequency.

The NEP can now be calculated according to equation (3.3). It is determined as NEP = 14 pW at a 1-kHz reference frequency and 1 second of measurement bandwidth. The specific detectivity is derived using equation (3.4), which gives $D^* = 7.14 \cdot 10^9 \frac{\text{cm}\sqrt{\text{Hz}}}{\text{W}}$. D^* allows a direct comparison of different detectors, as it is normalized to measurement bandwidth and detector area. Ultimately, the 'effective' D^*, which is defined in chapter 3, of a hybrid detector would have to compete with this value.

E. Atmospheric Transmission Impairments

In the following, the physical effects of the most common impairments of the atmosphere on a coherent beam are briefly discussed:

E.1. Atmospheric Scattering and Absorption

The atmosphere consists of molecules and particles (aerosols) of different shapes and sizes. These constituents lead to wavelength dependent scattering and absorption of light. As a consequence, the wavelength-dependent transmission decays with distance. In case of an inhomogeneous path through the atmosphere, e.g. with varying pressure and temperature, the integrated effect along the beam path must be determined. In the case of monochromatic radiation traversing an optical path length of ΔL on a homogeneous path, an exponential decay in intensity is observed according to the law of Lambert-Beer:

$$\tau = \exp\left(-\gamma \Delta L\right), \tag{E.1}$$

where $\gamma = \sigma + k$ is the attenuation coefficient. Here, σ is the scattering coefficient and k is the absorption coefficient. Both, σ and k depend on the size, type, and density distribution of molecules and particles in air:

$$\sigma = \sigma_m + \sigma_a \tag{E.2}$$
$$k = k_m + k_a,$$

where the index m denotes molecular absorption and scattering, and a the effects of aerosols.

The absorption and scattering coefficients due to aerosols with sizes comparable to the wavelength are usually smooth functions of wavelength and depend on the ratio of particle size to wavelength and the particle refractive index (Mie-scattering). Of course it is necessary to consider the particle distribution in this case. In case of molecular (Rayleigh-) scattering, the wavelength dependence is very nearly $\sigma_m \propto \lambda^{-4}$ [86]. Tabular values of k_a, σ_a and σ_m, taken from [86] at 0 km altitude, are plotted in figure E.1.

The molecular absorption coefficient, k_m, is of a different nature. The wavelength dependence is a highly oscillatory function, with a myriad of discrete absorption lines due to quantum mechanical transitions between different energy levels. The energies needed in order to excite such transitions depend on:

- the type of molecule

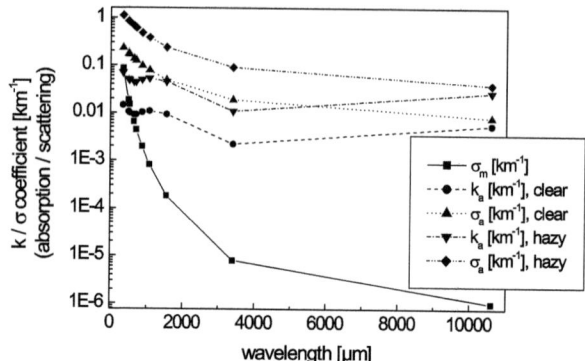

Figure E.1.: Measured molecular and aerosol scattering coefficients, as well as aerosol absorption coefficients, plotted as a function of wavelength up to 10 μm for two different visibility conditions (Taken from [86]).

- the type of interaction
- the harmonic order of the transition.

In addition, individual lines are broadened due to finite lifetimes of the excited states, as well as through Doppler- and pressure-broadening effects. Consequentially, an absorption spectrum with more or less pronounced absorption bands is observed in radiation transmitted through the atmosphere. Doppler broadening is dominant at low pressures below 0.1 atm [69]. In the lower 50 kilometers of the atmosphere, line broadening can be described to good accuracy by pressure broadening, and the molecular-absorption coefficient may be written as the Lorentzian function [86]:

$$k_m(\nu) = \frac{S\alpha}{\pi\left[(\nu - \nu_0^2 + \alpha^2\right]}, \qquad (E.3)$$

with line intensity S, line half width α, line center wavenumber ν_0 (in units cm^{-1}), and the wavelength at which the absorption coefficient is required ν. The line intensity and half-width depend on temperature and pressure. In case of a number of i contributing absorption lines at frequency ν, a summation according to

$$k_m = \sum_i k_{mi} \qquad (E.4)$$

becomes necessary.

In figure E.2, an example transmission spectrum through the atmosphere was calculated using SPECTRA [87] for a horizontal path of 2 km at sea level, assuming a mean-latitude USA summer atmospheric model. Here, only molecular absorption is considered, based on the HITRAN [69] and GEISA [96] databases. Obviously, there is a multitude of absorption lines forming complete bands, which are already smeared out due to an assumed instrument function with finite resolution of 10 cm^{-1} in wavenumbers. In the visible and near infrared up to 1 µm, there is naturally very little absorption at around 630 nm, 700 nm and 770 nm due to O_2 and few relatively weak absorption bands due to water vapor. In the near- to mid-infrared, water absorption bands are dominating, but also CO_2 molecules add strong absorption bands. Especially, the strong CO_2-band centered around 4.2-µm wavelength blocks radiation.

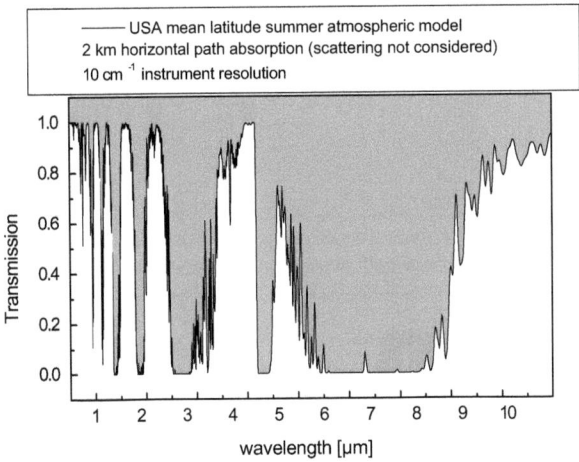

Figure E.2.: Atmospheric transmission on a horizontal path of 2 km in the atmosphere [87].

In-between these bands, transmission windows exist which are suitable for atmospheric transmission. The center wavelengths of these bands are listed in table E.1. Within the 5 and 8-µm wavelength band, transmission is blocked by water vapor, but another transmission window opens at around 10 µm.

Transmission bands, however, do not show the fine structure of absorption. Indeed, a narrow laser line may easily fit in-between individual absorption lines for data transmission purposes. This is shown in figure E.3 for the near- and mid-infrared ranges. A wavelength range where molecular absorption lines are nearly absent can be found at 3.82-µm wavelength, just in-between moderately strong water absorption lines. The next longer wavelength of equally high transmission is 4.06-µm, between the weak P and R branches of O_2 absorption; however, thermal background due to blackbody radiation from the earth is negligible only below 4-µm wavelength. Also, the transparency of $LiNbO_3$ falls off around 4 µm.

Center frequency	Adjacent absorption bands
750 nm	O_2 / H_2O
780 nm	O_2 / H_2O
870 nm	H_2O
1050 nm	H_2O
1240 nm	H_2O / O_2
1620 nm	H_2O / CH_4 / CO_2
2130 nm	H_2O / CO_2 / CH_4
2240 nm	H_2O / CH_4
4060 nm	H_2O / O_2

Table E.1.: Transmission bands in the near- to mid-infrared range [87].

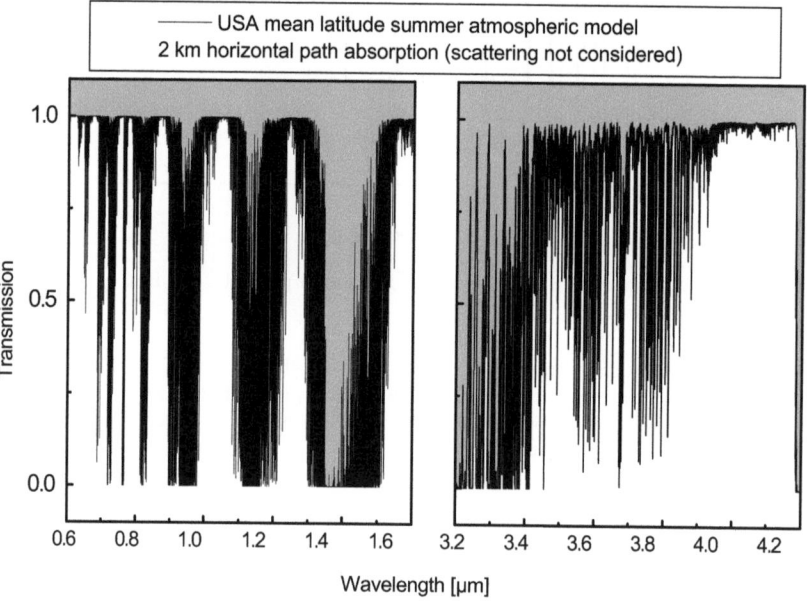

Figure E.3.: (Fairly) high resolution transmission spectra in the near- and mid-infrared ranges [87].

In addition to these considerations, the previously mentioned effect of scattering (Rayleigh and Mie) as well as absorption by aerosols must be considered. For this purpose, the data points in figure E.1 were fitted with analytic functions. Aerosol absorption was neglected in this case, due to oscillatory behavior. The resulting transmission curve is depicted in figure E.4

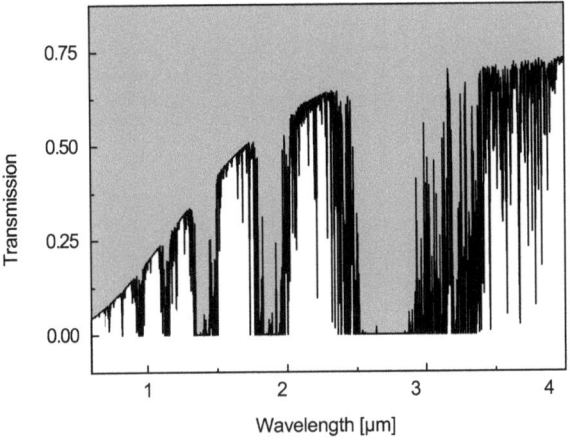

Figure E.4.: Transmission curve for a horizontal, 2-km path at 0-km altitude; US mid-latitude summer model, clear atmospheric conditions, considering molecular absorption and scattering, as well as aerosol scattering.

E.2. Beam Divergence and Spreading

Any optical beam (i.e. a beam emanating from an aperture of finite dimension) experiences divergence during propagation. From basic wave equations follows that a beam with Gaussian intensity distribution has the least possible divergence. In homogeneous media, a z-propagating beam is described by the complex field amplitude [97]

$$U(\rho, z) = A_0 \frac{W_o}{W(z)} \exp\left[-\frac{\rho^2}{W^2(z)}\right] \exp\left[-ikz - ik\frac{\rho^2}{2R(z)} + i\zeta(z)\right]. \quad (E.5)$$

Here, $W(z)$ and $R(z)$ are measures of beam half-width and curvature at a distance z from the beam waist with W_0, with the Rayleigh-range z_0 and the so-called Guoy-phase $\zeta(z)$. Time dependence is then described by the complex wavefunction $U(\vec{r}, t) = U(\rho, z) \exp(i\omega t)$, with angular frequency ω of the optical wave.

From E.5 follows the Gaussian beam intensity profile $I(\rho, z) = |U(\rho, z)|^2$:

$$I(\rho, z) = I_0 \left[\frac{W_0}{W(z)}\right]^2 \exp\left[-\frac{2\rho^2}{W^2(z)}\right]. \quad (E.6)$$

For $z \gg z_0$, a beam divergence half-angle of $\theta_0 = W_0/z_0$, or

$$\theta_0 = \frac{2}{\pi}\frac{\lambda}{2W_0}. \tag{E.7}$$

is derived. This is a well-known result which states that the divergence angle of a Gaussian beam is directly proportional to the wavelength divided by the beam waist. As a consequence, for equal transmission performance over large ranges, the aperture defining the waist diameter of a beam should scale proportional to the wavelength of light, so for MIR-transmission, a larger aperture is needed for collimation, compared to NIR or visible light.

E.3. Scintillation

The refractive index of air fluctuates with pressure and temperature. This leads to path deviation, irradiance-fluctuations as well as de-phasing (i.e. loss of coherence) of an optical beam traversing the turbulent atmosphere. This so-called scintillation is described assuming a spatial power spectrum of refractive-index fluctuations in a continuum of turbulent cells, leading to small-scale (causing diffraction) and large-scale (causing refraction) effects [98].

In order to describe the phenomenon, a scintillation index is defined as

$$\sigma_I^2 = \frac{\langle I^2 \rangle}{\langle I \rangle^2} - 1, \tag{E.8}$$

where the brackets denote a time (or ensemble) average. It is usually assumed that the irradiance can be expressed as the product $I = xy$, where x arises from large-scale turbulent eddies, and y arises from small-scale turbulent eddies. The variance of irradiance is then given by

$$\langle I^2 \rangle = \langle x^2 \rangle \langle y^2 \rangle = \left(1 + \sigma_x^2\right)\left(1 + \sigma_y^2\right) \tag{E.9}$$

with the normalized variances σ_x^2 and σ_y^2 of x and y, respectively. It was assumed that $\langle I \rangle = 1$. From equation E.9, the scintillation index (equation E.8) becomes

$$\sigma_I^2 = \left(1 + \sigma_x^2\right)\left(1 + \sigma_y^2\right) - 1 = \sigma_x^2 + \sigma_y^2 + \sigma_x^2\sigma_y^2. \tag{E.10}$$

The theory used to predict the scintillation index in equation E.10 is rather extensive, and care must to be taken to accurately describe the situation in weak and strong turbulence conditions. Considering a Gaussian beam, the phase-front radius of curvature and beam size will continually change due to diffraction as it propagates. Scintillation due to turbulence causes additional broadening of the spot size, focusing as well as de-focusing, and beam wander.

Andrews et al. derive the following expression for equation E.10 in case of a Gaussian beam [99]:

$$\sigma_I^2(\rho, z) = 4.42\sigma_1^2 \Lambda_e^{5/6} \frac{r^2}{W_e^2} + \exp\left[\frac{0.49\sigma_B^2}{\left(1+0.56\sigma_B^{12/5}\right)^{7/6}} + \frac{0.51\sigma_B^2}{\left(1+0.69\sigma_B^{12/5}\right)^{5/6}}\right] - 1 \quad (E.11)$$

at the radial and longitudinal position (ρ, z) in the beam, with the Rytov variance σ_1^2 of a plane wave, the Rytov variance σ_B^2 (σ_1^2) of a Gaussian beam, an effective spot radius W_e and an effective beam parameter $\Lambda_e = z/z_e$[1]. The authors neglected inner scale effects (i.e. scales at which turbulent energy dissipates) in order to reach this particular result.

In their formulation, propagation of a Gaussian beam is traced back the the Rytov variance $\sigma_1^2 = 1.23 C_n^2 k^{7/6} L^{11/6}$ for a plane wave, with the so-called refractive index structure constant C_n^2, wavenumber $k = 2\pi/\lambda$, and propagation distance L. The plane wave approximation had been well understood previously, however it could not accurately reproduce experimental data. Andrews et al. could show a good agreement between their theory represented by equation E.11 and experimental results, as well as numerical calculations performed by other research groups.

In the appendix to [81], calculations were performed by The Boeing Company associates using a basic spherical wave model which is much simpler to handle. The results are less exact due to simplification, however they expect the qualitative trends of the model to be the same. The scintillation index in this case is given by [98]

$$\sigma_I^2(\rho, z) = \exp\left[\frac{0.17\sigma_1^2}{\left(1+0.167\sigma_1^{12/5}\right)^{7/6}} + \frac{0.225\sigma_1^2}{\left(1+0.259\sigma_1^{12/5}\right)^{5/6}}\right] - 1, \quad (E.12)$$

which can be easily evaluated graphically for different wavelengths and propagation distances. I shall reproduce their findings here using some example calculations.

The refractive index structure parameter C_n^2 is a measure of turbulence. Values of $C_n^2 = 5 \cdot 10^{-14}$ m$^{-2/3}$, $C_n^2 = 5 \cdot 10^{-13}$ m$^{-2/3}$, and $C_n^2 = 5 \cdot 10^{-12}$ m$^{-2/3}$ are assumed values for weak, moderate, and strong turbulence, respectively [81]. Calculations have been performed for three wavelengths in the visible, NIR, and MIR (0.5 μm, 1.55 μm, and 3.8 μm, respectively) and results are shown in figure E.5.

In essence, the scintillation index increases to a peak as a function of σ_1, after which it declines again to a saturation value. Performance in the MIR is superior until the line crosses over with the 1.55-μm line. In a weakly turbulent environment, it crosses over at around 6.7 km; in a moderately turbulent environment at around 2 km; and in heavy turbulence at around 550 m (assuming a spherical wave). This means that the severity of turbulence must be taken into account, as well as the intended propagation distance

---
[1]W_e and Λ_e are expressions for the enlarged beam area which is painted by the fluctuating and wandering beam at the receiver aperture [99].

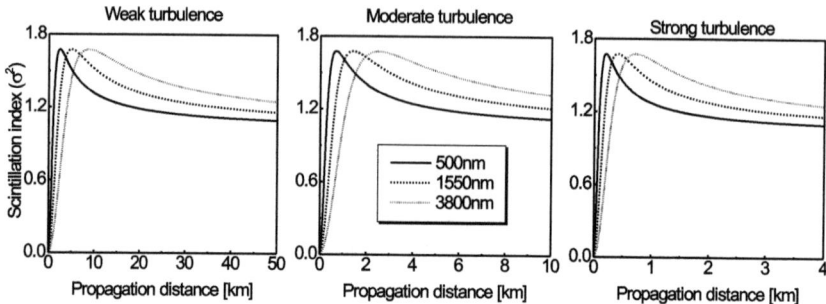

Figure E.5.: Scintillation index for weak, moderate, and strong turbulence conditions.

through turbulent areas, in order to estimate the performance of a given transmission system. In case that turbulent boundary layers of limited physical dimension should be traversed, use of MIR radiation definitely shows an advantage.

Bibliography

[1] T. H. Maiman. Stimulated Optical Radiation in Ruby. *Nature*, 187:493–494, 1960.

[2] A. L. Schawlow and C. H. Townes. Infrared and optical masers. *Phys. Rev.*, 112(6):1940–1949, Dec 1958.

[3] J. A. Armstrong, N. Bloembergen, J. Ducuing, and P. S. Pershan. Interactions between light waves in a nonlinear dielectric. *Phys. Rev.*, 127(6):1918–1939, Sep 1962.

[4] N. Bloembergen. Solid state infrared quantum counters. *Phys. Rev. Lett.*, 2(3):84–85, Feb 1959.

[5] D. H. Jundt, G. A. Magel, M. M. Fejer, and R. L. Byer. Periodically poled $LiNbO_3$ for high-efficiency second-harmonic generation. *Appl. Phys. Lett*, 59:2657–2659, 1991.

[6] Pantec Biosolutions AG Medical Laser. CTAN: http://www.pantec.com/index.php?src=gendocs&ref=ML_Main&category=Main.

[7] J. Kunsch, L. Mechold, A. Paraskevopoulos, G. Strasser, Ch. Mann, and Q. Yang. Spektroskopische Laserdioden und deren Zubehör. http://www.lasercomponents.com/fileadmin/user_upload/home/Datasheets/lc/applikationsreport/spektroskopische-laserdioden.pdf.

[8] N.S. Prasad. Optical communications in the mid-wave IR spectral band. *Free-Space Laser Communications*, pages 347–391, 2008.

[9] D. Hofmann. *Nichtlineare, integriert optische Frequenzkonverter für das mittlere Infrarot mit periodisch gepolten Ti:LiNbO$_3$-Streifenwellenleitern*. PhD thesis, Universität Paderborn, 2001.

[10] H. Herrmann. *Optisch nichtlineare Differenzfrequenzerzeugung abstimmbarer, kohaerenter Strahlung im mittleren Infrarotebereich in Ti:LiNbO$_3$-Streifenwellenleitern*. PhD thesis, Universität Gesamthochschule Paderborn, 1991.

[11] S. Orlov. *Integrated Optical Parametric Generators And Oscillators For The Mid-Infrared (MIR) Range*. PhD thesis, Universität Paderborn, 2008.

[12] D. Büchter. Generation of Mid-IR Radiation by Frequency Mixing in Ti:PPLN Waveguides for Environmental Analysis. Master's thesis, Universität Paderborn, 2007.

[13] M.-Ch. Wiegand. Absorptionsspektroskopie von Methangas mittels Differenz- und Summenfrequenzerzeugung in Ti:LiNbO$_3$-Wellenleitern. Bachelor's thesis, Universität Paderborn, 2008.

[14] Wikipedia article: Photonic integrated circuit. http://en.wikipedia.org/wiki/Photonic_integrated_circuit.

[15] W. Grundkötter. *Dynamik nichtlinearer Wechselwirkungen zweiter Ordnung in integriert optischen Wellenleitern*. PhD thesis, Universität Paderborn, 2007.

[16] E. Strake. Focus, 1991.

[17] B. Brecht. Focus revised, 2010.

[18] A. Yariv. *Optical Electronics*. Saunders College Publishing, Philadelphia, 4 edition, 1991.

[19] W. P. Huang. Coupled-mode theory for optical waveguides: an overview. *J. Opt. Soc. Am. A*, 11:963–983, 1994.

[20] F. Tian, C. Harizi, H. Herrmann, V. Reimann, R. Ricken, U. Rust, W. Sohler, F. Wehrmann, and S. Westenhofer. Polarization-independent integrated optical, acoustically tunable double-stage wavelength filter in LiNbO$_3$. *IEEE/OSA Journal of Lightwave Technology*, 12(7):1192–1197, 1994.

[21] H. Herrmann, K.D. Büchter, R. Ricken, and W. Sohler. Tunable Integrated Electro-Optic Wavelength Filter With Programmable Spectral Response. *Journal of Lightwave Technology*, 28(7):1051–1056, 2010.

[22] G. I. Stegeman and R. H. Stolen. Waveguides and fibers for nonlinear optics. *Journal of the Optical Society of America B*, 6(4):652–662, 1989.

[23] Rostislav V. Roussev, Carsten Langrock, Jonathan R. Kurz, and M. M. Fejer. Periodically poled lithium niobate waveguide sum-frequency generatorfor efficient single-photon detection at communication wavelengths. *Opt. Lett.*, 29(13):1518–1520, 2004.

[24] X. Liu, H. Zhang, and M. Zhang. Exact analytical solutions and their applications for interacting waves in quadratic nonlinear medium. *Optics Express*, 10(1):83–97, 2002.

[25] C. Canalias and V. Pasiskevicius. Mirrorless optical parametric oscillator. *Nature Photonics*, 1(8):459–462, 2007.

[26] V. Quiring. Personal correspondence, 2010.

[27] S. Orlov, W. Grundkötter, D. Hofmann, V. Quiring, R. Ricken, H. Suche, and W. Sohler. Mid-Infrared Integrated Optical Parametric Generators and Oscillators with Periodically Poled Ti:LiNbO$_3$ Waveguides. *Mid-Infrared Coherent Sources and Applications*, pages 377–392, 2008.

[28] R. Volk. *Lichtabsorption und optisch induzierte Brechungsindexänderungen in Ti:LiNbO$_3$-Streifenwellenleitern.* PhD thesis, Universität Gesamthochschule Paderborn, 1990.

[29] M. Asobe, O. Tadanaga, T. Yanagawa, T. Umeki, Y. Nishida, and H. Suzuki. Highpower mid-infrared wavelength generation using difference frequency generation in damage-resistant Zn:LiNbO$_3$ waveguide. *Electronics Letters*, 44(4):288–290, 2008.

[30] H. Hu, R. Ricken, and W. Sohler. Lithium niobate photonic wires. *Opt. Express*, 17(26):24261–24268, Dec 2009.

[31] Michele Gianella and Markus W. Sigrist. Infrared Spectroscopy on Smoke Produced by Cauterization of Animal Tissue. *Sensors*, 10(4):2694–2708, 2010.

[32] NovaWave Technologies, Inc. IRIS(TM) 1000 DFG-based Mid-infrared Laser System. http://www.novawavetech.com/products/photonics/mid-infrared-lasers/iris-1000-dfg-based-mid-infrared-laser-system.html.

[33] O. Tadanaga, T. Yanagawa, and M. Asobe. Special Feature: Light Source Technologies for Sensing. *NTT Technical Review*, 7, 2009.

[34] C. E. Webb and J. D. C. Jones. *Handbook of Laser Technology and Applications: Laser design and laser systems.* Taylor & Francis, 2004.

[35] Pranalytica High Power Laser System Model 1101-46-EGC. http://www.pranalytica.com/1101-46-EGC.html.

[36] Sacher Lasertechnik GmbH Quantum Cascade Laser Systems. http://www.sacher-laser.com/home/scientific-lasers/quantum_cascade_laser/quantum_cascade_lasers/quantum_cascade_laser_systems.html.

[37] Jerome Faist, Federico Capasso, Deborah L. Sivco, Carlo Sirtori, Albert L. Hutchinson, and Alfred Y. Cho. Quantum Cascade Laser. *Science*, 264(5158):553–556, 1994.

[38] Richard Maulini, Mattias Beck, Jérôme Faist, and Emilio Gini. Broadband tuning of external cavity bound-to-continuum quantum-cascade lasers. *Applied Physics Letters*, 84(10):1659–1661, 2004.

[39] J. Faist, F. Capasso, D.L. Sivco, A. L. Hutchinson, S. N. G. Chu, and A. Y. Cho. Short wavelength ($\lambda \sim$ 3.4 μm) quantum cascade laser based on strained compensated InGaAs/AlInAs. *Applied Physics Letters*, 72:680, 1998.

[40] I. Shoji, T. Kondo, A. Kitamoto, M. Shirane, and R. Ito. Absolute scale of second-order nonlinear-optical coefficients. *Journal of the Optical Society of America B*, 14(9):2268–2294, 1997.

[41] R. Regener and W. Sohler. Loss in low-finesse Ti: LiNbO$_3$ optical waveguide resonators. *Applied Physics B: Lasers and Optics*, 36(3):143–147, 1985.

[42] M. Serr. *Erzeugung von Pikosekunden-Lichtimpulsen hoher Energie und Repetitionsrate durch elektrooptisches Cavity-Dumping von diodengepumpten und passiv modengekoppelten Laseroszillatoren basierend auf Nd:YVO$_4$*. PhD thesis, TU Kaiserslautern, 2007.

[43] K.-D. F. Büchter, H. Herrmann, C. Langrock, M.M. Fejer, and W. Sohler. All-optical Ti:PPLN wavelength conversion modules for free-space optical transmission links in the mid-infrared. *Optics Letters*, 34(4):470–472, 2009.

[44] Klastech Senza 1064 nm. `http://www.klastech.com/senza-1064nm-laser_21.html`.

[45] XenICs XMID-394 InSb MIR camera. `http://www.xenics.com`.

[46] D. Büchter, H. Herrmann, C. Langrock, M. Fejer, and W. Sohler. Integrated optical PPLN transmitter and receiver modules for wavelength conversion of C-band signals to/from the mid infrared. In *34th European Conference on Optical Communication, 2008. ECOC 2008*, pages 1–2, 2008.

[47] D. H. Jundt. Temperature-dependent Sellmeier equation for the index of refraction, n_e, in congruent lithium niobate. *Opt. Lett.*, 22(20):1553–1555, 1997.

[48] Li Gui, Hui Hu, Miguel Garcia-Granda, and Wolfgang Sohler. Local periodic poling of ridges and ridge waveguides on X- and Y-Cut LiNbO$_3$ and its application for second harmonic generation. *Opt. Express*, 17(5):3923–3928, 2009.

[49] H. A. Haus. *Waves and fields in optoelectronics*, volume 1. Prentice-Hall Englewood Cliffs, NJ, 1984.

[50] O. M. Williams. A critique on the application of infrared photodetector theory. *Infrared Physics*, 26(3):141 – 153, 1986.

[51] F. Bordoni and A. D'Amico. Noise in sensors. *Sensors and Actuators A: Physical*, 21(1-3):17–24, 1990.

[52] W. Demtröder. *Laser Spectroscopy*. Springer Verlag Berlin Heidelberg, 1982.

[53] Hamamatsu.com. Hamamatsu InGaAs PIN Photodiodes. `http://www.sales.hamamatsu.com/en/products/solid-state-division/ingaas-pin-photodiodes/standard-type.php`.

[54] Femto Messtechnik GmbH OE-200-IN2. http://www.femto.de/.

[55] D. A. Kleinman. Theory of optical parametric noise. *Phys. Rev.*, 174(3):1027–1041, Oct 1968.

[56] Masaki Asobe, Osamu Tadanaga, Tsutomu Yanagawa, Hiroki Itoh, and Hiroyuki Suzuki. Reducing photorefractive effect in periodically poled ZnO- and MgO-doped LiNbO$_3$ wavelength converters. *Applied Physics Letters*, 78(21):3163–3165, 2001.

[57] A. Krier and Y. Mao. High performance uncooled InAsSbP/InGaAs photodiodes for the 1.8-3.4 µm wavelength range. *Infrared Physics & Technology*, 38(7):397 – 403, 1997.

[58] G. Temporão, S. Tanzilli, H. Zbinden, N. Gisin, T. Aellen, M. Giovannini, and J. Faist. Mid-infrared single-photon counting. *Optics letters*, 31(8):1094–1096, 2006.

[59] Lijun Ma, Oliver Slattery, and Xiao Tang. Experimental study of high sensitivity infrared spectrometer with waveguide-based up-conversion detector. *Opt. Express*, 17(16):14395–14404, 2009.

[60] W. Demtröder, M. Keil, Th. Platz, and H. Wenz. Advances of sensitive techniques in laser spectroscopy. *Spectrochimica Acta Part B: Atomic Spectroscopy*, 63(2):183 – 190, 2008. Honoring Issue A Collection of Papers on Atomic, Molecular and Laser Spectroscopy Dedicated to James D. Winefordner.

[61] M. Falk and E. Whalley. Infrared spectra of methanol and deuterated methanols in gas, liquid, and solid phases. *The Journal of Chemical Physics*, 34:1554, 1961.

[62] A. E. Klingbeil, J. B. Jeffries, and R. K. Hanson. Tunable mid-IR laser absorption sensor for time-resolved hydrocarbon fuel measurements. *Proceedings of the Combustion Institute*, 31(1):807–815, 2007.

[63] C. Frankenberg, JF Meirink, P. Bergamaschi, APH Goede, M. Heimann, S. Körner, U. Platt, M. Van Weele, and T. Wagner. Satellite chartography of atmospheric methane from SCIAMACHY on board ENVISAT: Analysis of the years 2003 and 2004. *Journal of Geophysical Research*, 111(D7):D07303, 2006.

[64] B. Culshaw, G. Stewart, F. Dong, C. Tandy, and D. Moodie. Fibre optic techniques for remote spectroscopic methane detection–from concept to system realisation. *Sensors and actuators B: chemical*, 51(1-3):25–37, 1998.

[65] D. W. Lachenmeier. Rapid quality control of spirit drinks and beer using multivariate data analysis of Fourier transform infrared spectra. *Food Chemistry*, 101(2):825–832, 2007.

[66] E. T. S. Skibsted, J. A. Westerhuis, A. K. Smilde, and D. T. Witte. Examples of NIR based real time release in tablet manufacturing. *Journal of pharmaceutical and biomedical analysis*, 43(4):1297–1305, 2007. Cited By (since 1996): 7.

[67] A. Migus, A. Antonetti, J. Etchepare, D. Hulin, and A. Orszag. Femtosecond spectroscopy with high-power tunable optical pulses. *Journal of the Optical Society of America B*, 2(4):584–594, 1985.

[68] P. Weibring, D. Richter, J. G. Walega, and A. Fried. First demonstration of a high performance difference frequency spectrometer on airborne platforms. *Optics Express*, 15(21):13476–13495, 2007.

[69] L. S. Rothman, I. E. Gordon, A. Barbe, D. C. Benner, P. F. Bernath, M. Birk, V. Boudon, L. R. Brown, A. Campargue, J. P. Champion, et al. The HITRAN 2008 molecular spectroscopic database. *Journal of Quantitative Spectroscopy and Radiative Transfer*, 110(9-10):533–572, 2009.

[70] K.-D. Büchter, M.-C. Wiegand, H. Herrmann, and W. Sohler. Nonlinear optical down- and up-conversion in PPLN waveguides for mid-infrared spectroscopy. In *Lasers and Electro-Optics 2009 and the European Quantum Electronics Conference. CLEO Europe - EQEC 2009. European Conference on*, pages 1 –1, jun. 2009.

[71] K. P. Petrov, S. Waltman, U. Simon, R. F. Curl, F. K. Tittel, E. J. Dlugokencky, and L. Hollberg. Detection of methane in air using diode-laser pumped difference-frequency generation near 3.2 μm. *Applied Physics B: Lasers and Optics*, 61:553–558, 1995. 10.1007/BF01091213.

[72] Ch. K. Kao. Nobel prize, 2009. http://nobelprize.org/nobel_prizes/physics/laureates/2009/.

[73] IEEE 802.11 WLAN standard. http://grouper.ieee.org/groups/802/11/.

[74] W. T. Buttler, R. J. Hughes, P. G. Kwiat, S. K. Lamoreaux, G. G. Luther, G. L. Morgan, J. E. Nordholt, C. G. Peterson, and C. M. Simmons. Practical Free-Space Quantum Key Distribution over 1 km. *Phys. Rev. Lett.*, 81(15):3283–3286, Oct 1998.

[75] LightPointe, Inc. Free Space Optics Topologies. http://www.freespaceoptics.org/freespaceoptics/topologies/default.cfm.

[76] V. W. S. Chan. Free-space optical communications. *Lightwave Technology, Journal of*, 24(12):4750–4762, 2006.

[77] Canobeam, Canon Inc. http://www.freespaceoptics.com/index.php?page=productinfo, 2010.

[78] fSONA Optical Wireless, fSONA Networks Corp. http://www.fsona.com/product.php?sec=models_overview, 2010.

[79] T. Aellen, M. Giovannini, J. Faist, and J. P. von der Weid. Feasibility Study of Free-Space Quantum Key Distribution in the Mid-infrared. *Quantum Information and Computation*, 8(1&2):0001–0011, 2008.

[80] International Telecommunication Union. DWDM Dense Wavelength Division Multiplexing. http://www.itu.int/dms_pub/itu-t/oth/1D/01/T1D010000090001PDFE.pdf.

[81] P. S. Cho, G. Harston, K. D. F. Büchter, D. Soreide, J. M. Saint Clair, W. Sohler, Y. Achiam, and I. Shpantzer. Optical homodyne RZ-QPSK transmission through wind tunnel at 3.8 and 1.55 μm via wavelength conversion. In *Proc. of SPIE Vol*, volume 7324, pages 73240A–1, 2009.

[82] W. O. Popoola and Z. Ghassemlooy. BPSK subcarrier intensity modulated free-space optical communications in atmospheric turbulence. *Lightwave Technology, Journal of*, 27(8):967–973, 2009.

[83] U.S.S. Atmosphere. US Standard Atmosphere, 1976. *US Government Printing Office, Washington, DC*, 1976.

[84] ICAO. Manual of the ICAO Standard Atmosphere: extended to 80 kilometres (262 500 feet). http://www.icao.int/cgi/ISBN_txt.pl?16, 1993.

[85] International Telecommunications Union. Attenuation by Atmospheric Gases. *ITU*, pages 676–6, 2005.

[86] W. G. Driscoll and W. Vaughan. *Handbook of Optics*. McGraw-Hill, 1978.

[87] Institute of Atmospheric Optics. http://spectra.iao.ru/en/, 2010.

[88] R. Martini, C. Gmachl, J. Falciglia, F.G. Curti, C. G. Bethea, F. Capasso, E. A. Whittaker, R. Paiella, A. Tredicucci, A. L. Hutchinson, et al. High-speed modulation and free-space optical audio/videotransmission using quantum cascade lasers. *Electronics Letters*, 37(3):191–193, 2001.

[89] E. Ip, D. Büchter, C. Langrock, J. M. Kahn, H. Herrmann, W. Sohler, and M. M. Fejer. QPSK transmission over free-space link at 3.8 μm using coherent detection with wavelength conversion. In *Optical Communication, 2008. ECOC 2008. 34th European Conference on*, pages 1–2. IEEE, 2008.

[90] C. Langrock, S. Kumar, J. E. McGeehan, A. E. Willner, and M. M. Fejer. All-optical signal processing using $\chi^{(2)}$ nonlinearities in guided-wave devices. *Lightwave Technology, Journal of*, 24(7):2579–2592, 2006.

[91] H. Hu, R. Nouroozi, R. Ludwig, B. Huettl, C. Schmidt-Langhorst, W. Sohler, and C. Schubert. Polarization-insensitive all-optical wavelength conversion of 320 Gb/s RZ-DQPSK signals using a Ti:PPLN waveguide. *Applied Physics B: Lasers and Optics*, pages 1–8, 2010.

[92] D. Richter, A. Fried, and P. Weibring. Difference frequency generation laser based spectrometers. *Laser & Photonics Reviews*, 3(4):343–354, 2009.

[93] Nufern.com. Nufern 1060XP fiber. http://www.nufern.com/fiber_detail.php/16.

[94] OEC GmbH MCT detector. http://www.femto.de/.

[95] AMETEK Inc. *Model 7265 Dual Phase DSP Lock-in Amplifier*.

[96] N. Jacquinet-Husson, E. Arie, J. Ballard, A. Barbe, G. Bjoraker, B. Bonnet, L. R. Brown, C. Camy-Peyret, J. P. Champion, A. Chedin, et al. The 1997 spectroscopic GEISA databank. *Journal of Quantitative Spectroscopy and Radiative Transfer*, 62(2):205, 1999.

[97] B. E. A. Saleh and M. C. Teich. *Fundamentals of photonics*. New York: Wiley, 1991.

[98] L. C. Andrews, R. L. Phillips, C. Y. Hopen, and M. A. Al-Habash. Theory of optical scintillation. *JOSA A*, 16(6):1417–1429, 1999.

[99] L. C. Andrews, M. A. Al-Habash, C. Y. Hopen, and R. L. Phillips. Theory of optical scintillation: Gaussian-beam wave model. *Waves in Random and Complex Media*, 11(3):271–291, 2001.

[100] C. Langrock. Carsten Langrock's Homepage. http://www.stanford.edu/~langrock/.

Danksagung

Zunächst möchte ich mich herzlich bei Prof. Dr. Wolfgang Sohler für seine Unterstützung und Anleitung bedanken, die die Fertigstellung dieser Arbeit ermöglicht hat. Insbesondere bedanke ich mich auch für die bereichernden Auslandsaufenthalte in Stanford und Triest, die mir im Rahmen meiner Tätigkeiten ermöglicht wurden.

Vielen Dank an Viktor Quiring und Raimund Ricken, die zusammen mit ihren jeweiligen Auszubildenden viel Zeit und Mühe in die Herstellung hochwertiger Wellenleiterproben und anderes gesteckt haben. Danke auch an Irmgard Zimmermann für Ihre unkomplizierte Art und Unterstützung bei allen nicht-wissenschaftlichen Aufgaben. Nicht zuletzt gilt mein Dank Herrn Hartung und seinem Team in der Mechanikwerkstatt, welche mir häufig bei der sehr kurzfristigen Herstellung von Mechanikkomponenten entgegengekommen sind. Dr. Harald Herrmann und Dr. Hubertus Suche danke ich für die kontinuierliche Unterstützung und hilfreiche Diskussionen.

Vielen Dank an die lieben Kollegen, die ich im Laufe meiner Zeit in der AG Sohler kennengelernt habe und mit denen ich schöne Erinnerungen verbinde. Dazu gehören Dr. Miguel Garcia-Granda, Matthew George, Dr. Li Gui, Ansgar Hellwig, Dr. Hui Hu und seine liebe Frau, Dr. Rahman Nouroozi und seine Familie, Dr. Sergey Orlov, Abu Thomas und Marie-Christin Wiegand. Sie haben alle beigetragen zu einem sehr angenehmen Arbeitsklima in der Integrierten Optik.

Lieben Dank auch an Prof. Dr. Christine Silberhorn für die freundliche Zusammenarbeit. Auch Ihren Mitarbeitern ein Dankeschön für die angenehme Zeit in der Übergangsphase des Lehrstuhls. Danke an Benjamin Brecht für die Windows Version von FOCUS, die sich noch als sehr praktisch erwiesen hat. Auch bedanke ich mich für die weitere Unterstützung von Prof. Dr. Cedrik Meier, der mich in der finalen Phase meiner Promotion kurzfristig in seine Arbeitsgruppe aufgenommen hat.

Die nette Zeit in Stanford in der Arbeitsgruppe von Prof. Martin Fejer soll nicht unerwähnt bleiben. Die Zusammenarbeit mit Carsten Langrock, Ph.D.[2] und Ezra Ip, Ph.D. in Stanford, und Mitarbeitern von CeLight, Inc. und The Boeing Company in Seattle war sehr unkompliziert und spannend.

Kai-Daniel Frank Büchter, Januar 2011

---

[2]Ich bin so frei und zitiere: "Dr. Langrock for you, of course, unless we are in Germany, in which case I would get sued and thrown into prison for impersonating a Dr." [100]

Die VDM Verlagsservicegesellschaft sucht für wissenschaftliche Verlage abgeschlossene und herausragende

Dissertationen, Habilitationen, Diplomarbeiten, Master Theses, Magisterarbeiten usw.

für die kostenlose Publikation als Fachbuch.

Sie verfügen über eine Arbeit, die hohen inhaltlichen und formalen Ansprüchen genügt, und haben Interesse an einer honorarvergüteten Publikation?

Dann senden Sie bitte erste Informationen über sich und Ihre Arbeit per Email an *info@vdm-vsg.de*.

Sie erhalten kurzfristig unser Feedback!

VDM Verlagsservicegesellschaft mbH
Dudweiler Landstr. 99
D - 66123 Saarbrücken

Telefon +49 681 3720 174
Fax +49 681 3720 1749

www.vdm-vsg.de

Die VDM Verlagsservicegesellschaft mbH vertritt

Printed by Books on Demand GmbH, Norderstedt / Germany